Quality Protein Maize

Edited by

Edwin T. Mertz

Purdue University
West Lafayette, Indiana

The American Association of Cereal Chemists
St. Paul, Minnesota, USA

This publication is based primarily on presentations from the Quality
Protein Maize Symposium at the 75th Annual Meeting of the American
Association of Cereal Chemists held in Dallas, Texas, on October 14–18,
1990. To make the information available in a timely and economical
fashion, this book has been reproduced directly from typewritten copy
submitted in final form to the American Association of Cereal Chemists
by the editor of this volume. No editing or proofreading has been done
by the Association.

Library of Congress Catalog Card Number: 92-71401
International Standard Book Number: 0-913250-75-9

Reference in this volume to a company or product name by personnel of
the U.S. Department of Agriculture or anyone else is intended for explicit
description only and does not imply approval or recommendation of the
product to the exclusion of others that may be suitable.

Printed in the United States of America on acid-free paper

American Association of Cereal Chemists
3340 Pilot Knob Road
St. Paul, MN 55121-2097, USA

PREFACE

Dr. Lloyd Rooney organized and chaired a symposium on quality protein maize for the Protein Division of the American Association of Cereal Chemists in Dallas, Texas, October 17, 1990. The articles in this book are based primarily on that symposium with additional articles by Dr. Glover, Dr. Magnavaca and myself.

When Dr. Rooney invited me to serve as editor, I was pleased to accept the assignment. Editing in this area was not a new task for me. Twenty-five years ago I was co-editor of the Proceedings of the First High Lysine Corn Conference which was held at Purdue[1], and 16 years ago I was co-editor of the Proceedings of the Second High Lysine Corn Conference held at the International Maize and Wheat Improvement Center in El Batan, Mexico[2].

The papers presented in this volume show that there has been a resurgence of interest in quality protein maize. With the vastly improved germplasm now available, QPM should be much more acceptable to maize growers both in developing and developed countries. The full potential of QPM will be realized when it replaces normal maize as a human food and an animal feed in all areas of the world.

Edwin T. Mertz
Department of Agronomy
Purdue University
West Lafayette, IN

---

1. Proc. High Lysine Corn Conf. 1966. Corn Refiners Association, Washington, D.C.
2. High-Quality Protein Maize. 1975. Dowden, Hutchinson and Ross, Stroudsburg, PA.

ACKNOWLEDGEMENT

The editor wishes to thank Mrs. Karen Clymer, Dept. of Agronomy, Purdue University, for preparing camera-ready copy of five chapters in this book. Her work was supported by the Purdue Sorghum Nutritional Quality Project (J.D. Axtell and G. Ejeta), INTSORMIL Project, DAN-1254-G-00-002100, Agency for International Development, Washington, D.C.

TABLE OF CONTENTS

TABLE OF CONTENTS *(Continued)*

TABLE OF CONTENTS *(Continued)*

DISCOVERY OF HIGH LYSINE, HIGH TRYPTOPHAN CEREALS

Edwin T. Mertz
Department of Agronomy, Purdue University
West Lafayette, IN 47907

When I arrived at Purdue University in 1946 as an Assistant Professor in the Department of Biochemistry, I was invited to become a member of the Utilization Research Committee that had just been formed in the Agricultural Experiment Station to find new uses for the cereal grain surpluses then piling up in government storage facilities. I decided to start a research project on the major crop in Indiana - corn (*Zea mays*). My objective was to improve the nutritional value of corn proteins, which had to be combined with more expensive supplements such as meat, milk, eggs, whole legumes, or soybean meal in both animal and human diets.

My students and I began a search for a corn kernel that had a lower content of zein (the major protein in corn), and higher levels of other more nutritious proteins (i.e. albumins, globulins and glutelins). The first screening for a corn variety with these qualities was done by my student Ricardo Bressani, who has contributed a chapter to this book. It was continued by a graduate student from Ohio, Lynn Bates. We analyzed corn endosperms (seed minus the seed coat and embryo) at first for zein. We expected varieties with low levels of zein to have better protein quality, since zein is almost devoid of the two essential amino acids, lysine and tryptophan. In the early sixties we obtained one of the first automatic amino acid analyzers and then looked only for higher levels of lysine in the endosperm.

From 1957 to 1961, Dr. H.H. Kramer, Department of Agronomy, sent over corn varieties, both normal and mutants for our screening program. None were unusually high in lysine. When he left the university in 1961, Dr. Oliver E. Nelson, Department of Botany and Plant Pathology, took his place and supplied us with samples for analysis. The first samples from Dr. Nelson came from the Mexican Seed Bank and represented many races from Central and South America. None of these were high in lysine. Dr. Nelson was aware of the fact that we had found a lower level of lysine (grams lysine per 100 grams of protein) in the variety Illinois HP [high (18%) protein] than in the variety Illinois LP [low (4%) protein]. He was impressed with the fact that the Illinois HP was much more vitreous than the Illinois LP, which had a floury or chalky endosperm. Thinking that flouriness might be associated with high lysine, he sent four floury mutant types of kernels, floury-1, floury-2, opaque-1, and opaque-2 to us for analysis.

<u>Discovery of the high lysine level in opaque-2 maize</u>. On November 18, 1963, the endosperms of these samples were analyzed by Lynn Bates. Floury-1 and Opaque-1 had 2 grams of lysine (the normal level found in corn), but floury-2 contained 3 grams and opaque-2 contained 4 grams of lysine per 100 grams of protein. Since two of the samples of floury kernels (floury-1 and opaque-1) have normal lysine levels, flouriness is not closely linked to high lysine. Nevertheless, Dr. Nelson's thought that it might be linked paid big dividends. The opaque-2 mutant (now called quality protein maize or high lysine corn) still ranks at the top of all viable single mutants in its level of lysine, although many other high lysine types have been found (Mertz 1986). The opaque-2 endosperm also contains a much higher level of tryptophan.

After this discovery, plant breeders searched for similar mutations in other economically important cereal grains. Two high lysine types of barley were identified. The first, a naturally occurring mutation in Ethiopia was identified by Dr. Lars Munck, in

Sweden in 1968, and called HyProly. A chemically
induced mutation (1508) was identified by Ingversen
and coworkers about five years later. In the early
seventies Singh and Axtell identified high lysine
mutants of sorghum in germplasm from Ethiopia; and a
short time later Mohan and Axtell isolated chemically
induced high lysine mutants from temperate germplasm.
No high lysine mutants have been identified in the
other economically important cereals. Both opaque-2
maize and high lysine sorghum (Hassen et al. 1986)
have increased levels of tryptophan, but the high-
lysine gene(s) in barley do not affect the tryptophan
level (Newman et al. 1990). Levels of the latter are
adequate in both normal sorghum and normal barley.

<u>Nutritional effect of the opaque-2 gene</u>. In 1965,
feeding tests with young rats showed that animals
receiving only opaque-2 maize, minerals and vitamins
gained four times faster than those receiving normal
maize (Mertz et al. 1965). This was repeated with
weanling pigs (Cromwell et al. 1967) with the same
results. In Guatemala, Dr. Bressani showed that
opaque-2 maize had 90% of the nutritive value of milk
proteins in young children (Bressani 1966). In
Colombia, Doctors Harpstead, and Pradilla showed that
young children who had developed a severe protein
deficiency disease (kwashiorkor) on a diet of normal
maize could be brought back to health on a diet
containing only opaque-2 maize as the source of
protein (Harpstead 1971).
    There was great hope during the late 1960s that
opaque-2 would improve the health of millions of
people in the countries where corn was a major staple
in the diet. Unfortunately, this did not occur. The
soft endosperm opaque-2 grain had many undesirable
characteristics: a chalky, not shiny kernel, reduced
yields, lower resistance to fungi and insects in the
field and in storage, and a longer drying time. Most
of the agricultural research centers and hybrid corn
companies abandoned their research programs on opaque-
2 maize in the early 1970s.

3

<u>Three research centers that persisted</u>. Despite almost
universal abandonment, research to improve the
agronomic qualities continued at three locations, the
International Maize and Wheat Improvement Center
(CIMMYT) near Mexico City (see Dr. Villegas' chapter),
the University of Natal, South Africa (see Dr. Gever's
chapter), and Crow's Hybrid Corn Company in Milford,
Illinois. For more than a decade, maize breeders and
cereal chemists at these three locations continued to
do research with the opaque-2 gene. At CIMMYT and in
S. Africa they wanted to introduce gene modifiers that
would change the soft, starchy, chalky endosperm to a
vitreous type preferred by farmers in tropical
countries. At Crow's Hybrid Corn Company they wanted
to produce a high yielding midwest hybrid with
excellent harvesting and storage qualities and a soft,
starchy endosperm easily chewed and digested by hogs
and beef cattle. By the early 1980s all three
research centers had reached their objective.

CIMMYT has been distributing white and yellow
flint and dent open-pollinated opaque-2 varieties with
a hard endosperm, to national maize programs for many
years. The national programs have used the CIMMYT
germplasm to develop their own locally adapted open-
pollinated types. Some are testing hybrids made from
the synthetic varieties. CIMMYT has called their hard
endosperm opaque-2 germplasm quality protein maize
(QPM) to distinguish it from the soft endosperm
varieties which they call QPM-SE. Four QPM open-
pollinated varieties already released commercially
are: <u>Nutricta</u> in Guatemala, <u>Nutri-Guarani-V241</u> in
Paraguay, <u>Tuxpeno 102</u> in China, and <u>Population 63</u> in
Vietnam.

In 1983, the National Center for Research on Corn
and Sorghum (CNPMS) in Sete Lagoas, Minas Gerais,
Brazil, introduced 23 varieties of CIMMYT QPM. After
5 years of research, a white-colored variety with
sufficient adaptation to the diverse regions of the
country and of high nutritional value was developed.
It was named BR 451 and has 85% more lysine and
tryptophan than normal corn (The Sete Lagoas High
Nutrition Corn, 1989). Brazil has a strong program in

place at present to hasten acceptance of their
varieties by Brazilian farmers (see Dr. Magnavaca's
chapter), and a project has been initiated in Ghana to
stimulate adoption of CIMMYT QPM by the farmers (see
Dr. Borlaug's chapter).

Crow's Hybrid Corn Company achieved its
objectives by having available for the past decade,
high yielding varieties of soft endosperm opaque-2
hybrids (Hi-Lysine Corn) adapted to the short, medium
and long summers in the midwest. It is estimated that
farmers in the midwest produce one million tons of
Crow's Hi-Lysine corn annually, all of which stays on
the farm to be fed to the livestock, mainly growing
and finishing pigs.

Studies at the University of Nebraska (Asche et
al., 1985) showed that when Crow's Hi-Lysine corn was
fed in place of normal corn to all classes of pigs,
from weaning to finishing, the total level of dietary
protein supplement (usually soybean meal) could be
reduced by 2 percent. For example, in a 16% protein
ration, commonly used for fast growing pigs, the
soybean oil meal could be reduced 23 percent. This is
a significant savings, since soybean meal usually
costs two to three times the cost of corn.

Need for quality protein in the human diet. In 1973,
the United Nations FAO/WHO Report on Human
Requirements for Protein and Calories substantially
lowered the previous protein requirement figures.
Some nutritionists then said that malnutrition was due
only to a lack of calories, not protein. The emphasis
was, therefore, placed on increasing starch yields
rather than improving the quality of the protein.
However, late in 1987, a research team from three U.S.
universities (V. Young, D. Bier and P. Pellett)
reported that the human requirement for lysine,
leucine, valine and threonine was two to three times
higher than that proposed by FAO/WHO in 1973 and again
in 1985. For example, the adult needs a protein with
the same level of lysine (about 5 grams of lysine per
100 grams of protein) previously recommended by
FAO/WHO for the 2 year old child. This new finding

has been accepted by FAO and implies that adults would also benefit greatly from QPM because of its higher level of lysine and tryptophan (V. Young et al. 1991).

Research in Lima, Peru by Dr. George Graham and coworkers has shown that babies in their second year of life grow normally when QPM is fed as the only source of protein in the diet (Graham et al., 1990). In this study, ten infants were fed QPM as the only source of protein and fat which supplied 90 percent of the calories. Ten percent of the calories came from sugar. A standard milk formula diet was fed to a similar control group of 10 children at the same calorie level. The diet containing QPM supported growth rates that were <u>statistically indistinguishable</u> from those found in the children receiving the milk formula.

In another study showing the superior digestibility and utilization of QPM in recovering malnourished infants, Dr. Graham (1990) concluded that: "To anyone familiar with the nutritional problems of weaned infants and small children in the developing world, and with the fact that millions of them depend on maize for most of their dietary energy, nitrogen and essential amino acids, the potential advantages of QPM are enormous. To assume that these children will always be given a complementary source of nitrogen and amino acids is a cruel delusion."

<u>Conclusion</u>. The weight of evidence that has accumulated over the past 27 years regarding the superior protein quality of hard and soft endosperm opaque-2 maize should serve as a strong motivation to the governments and Ministries of Health and Ministries of Agriculture in developing countries to vigorously sponsor national production of QPM in those countries where maize makes up a major part of the human diet. Hard or soft endosperm opaque-2 maize should also replace normal maize as a feed grain in the U.S. and other maize producing countries to reduce the cost of animal protein. The National Academy of Sciences considers QPM to be an "underexploited crop" (Quality Protein Maize, 1988) that should be more

widely used in both developed and developing countries.

LITERATURE

Asche, G.L., Lewis, A.J., Peo, Jr., E.R. and Crenshaw, J.D. 1985. The nutritional value of normal and high lysine corns for weanling and growing-finishing swine when fed at four lysine levels. J. Animal Sci. 60:1412-1428.

Bressani, R. 1966. Protein quality of opaque-2 maize in children. Proc. High Lysine Corn Conf. Corn Industries Res. Foundation. Washington, D.C. pp. 34-39.

Cromwell, G.L., Pickett, R.A. and Beeson, W.M. 1967. Nutritional value of opaque-2 corn for swine. J. Animal Sci. 26:1325-1331.

Graham, G.G., Lembcke, J. and Morales, E. 1990. Quality-Protein Maize as the Sole Source of Dietary Protein and Fat for Rapidly Growing Young Children. Pediatrics 85(1):85-91.

Graham, G.G., Lembcke, J., Lancho, E. and Morales, E. 1990. Quality Protein Maize's Digestibility and Utilization by Recovering Malnourished Infants. Pediatrics 83(3):416-421.

Hassen, M.M., Mertz, E.T., Kirleis, A.W, Ejeta, G., Axtell, J.D. and Villegas, E. 1986. Tryptophan levels in normal and high-lysine sorghums. Cereal Chem. 63:175.

Harpstead, D.D. 1971. High lysine corn. Sci. American 225(2):34-42.

Mertz, E.T. 1986. Genetic and biochemical control of grain protein synthesis in normal and high lysine cereals. World Review of Nutrition and Dietetics 48:222-262.

Mertz, E.T., Veron, O, Bates, L.S., and Nelson, O.E. 1965. Growth of rats fed on opaque-2 maize. Science 148:1741-1742.

Newman, C.W., Overland, M., Newman, R.K., Bang-Olsen, K, and Pederson, B. 1990. Protein quality of a new high-lysine barley derived from R150-1508. Can. J. Anim. Sci. 70:279-285.

Quality Protein Maize. 1988. NAS Monograph. National Academy Press. Washington, D.C.

The Sete Lagoas High Nutrition Corn. 1989. National Center for Research on Corn and Sorghum (CNPMS). Sete Lagoas, Minas Gerais, Brazil.

Young, V.R., Bier, D.M., and Pellett, P.L.. 1989. A theoretical basis for increasing current estimates of the amino acid requirements in adult man, with experimental support. Am. J. Clin. Nutr. 50:80-92.

Young, V.R. and Pellet, P. 1991. Human amino acid requirements with reference to protein quality. Proc. International Conf. on Sorghum Nutritional Quality. Purdue University Press, W. Lafayette, IN.

CORN PROTEIN -- GENETICS, BREEDING, AND VALUE IN FOODS AND FEEDS*

David V. Glover
Dept. of Agronomy
Purdue University
West Lafayette, IN 47907

There is increasing talk of opportunities for corn producers to specialize in particular products that have high return for the end user and which may command a premium. This suggests that there may be greater emphasis on corn that will be genetically tailored for the food and non-food processor, and the consumer. The result will be "non-commodity" differentiated corns.

Industries needs for corn are becoming increasingly segmented, opening new markets for corn that has specialized characteristics. The grain processing industry has very specific demands: both wet millers and dry millers require special traits. Does that mean that corn breeders can afford to place more emphasis on specialty hybrids? In perspective, we should remember that specialty corn markets are relatively small. The development of hybrids for specialty uses has been prohibitive in the past because of either cost or difficulty. Improvements in corn for specific protein modifications are no exception. The problem in the past has been convincing end-users that specific value-added genetic qualities are worth paying more for. The livestock industry, for example, already the largest user of corn, has a potential for value-added nutritionally improved corn. However, the industry's main interest in corn is for its energy value, and that comes primarily from starch. The recent increase in crop

---

*Reprinted from the Proceedings of the 43rd Annual Corn and Sorghum Research Conference, 1988 with permission of the author and publishers. This is part of a paper entitled "Corn Protein and Starch -- Genetics, Breeding, and Value in Foods and Feeds".

biotechnology research in food and non-food industries
is building momentum for utilization or value-added
genetics. Here, the orientation is on the end use of
agricultural raw materials, with a focus on
added value and higher return. It is implied that
there may be significant opportunities to extend the
value-added food or non-food industry component back
to the farm. Regardless of what may come to pass, the
issue remains, will the markets for improvements
develop, and which specific genes or traits are to be
selected as economically viable targets?

Most of my comments on corn protein are taken from
our recent review (Glover and Mertz, 1987). Globally,
about two-thirds of total maize production is used for
livestock feed. It is an important food material
which supplies 19 percent of the world's food
calories. Corn also contributes some 42 million tons
of protein a year, which represents 15 percent of the
world's annual production of food-crop protein.
Understandably high grain yields are the objectives of
most corn improvement programs, and breeding for
nutritional quality or special purposes is often of
minor importance, simply because of a lack of
incentives to do so. A part of this stems from
uncertainty of future demands for special-purpose
corn. Furthermore, breeding of these special types
has generally resulted in major shifts in grain
quantity as emphasis was placed on quality
improvement.

Protein Quantity

Protein quantity in corn grain is a function of
cultural practices and heredity. The current average
protein content of U.S. hybrids ranges between 9 and
11% (moisture-free basis). Selection for high protein
in the classical "Illinois Chemicals" experiments
increased protein in both germ and endosperm. The
Illinois High Protein strain has about 27% protein by
weight, but protein yield is not improved in
proportion. The increase in endosperm protein was
associated with increase in the zein (prolamin)
fraction so the quality does not increase in
proportion to quantity. There is a strong negative

correlation between protein and yield. Given the
right selection indices, protein values may be
increased one to three percentage points along with
improved grain yield (Kauffmann and Dudley, 1979).
Currently, not much interest exists in developing
hybrids with higher protein potential because, in the
U. S., soybean protein is economically available.

Genetic Modification of Endosperm Storage Proteins

The storage proteins of corn comprise a group of
alcohol-soluble proteins called prolamins (zeins).
These proteins are noted for their high content of
proline and glutamine and neither lysine nor
tryptophan has been found in any of the several zeins
whose sequences are known. Since prolamins account
for approximately 50% of the total protein in the
endosperm, corn is deficient in lysine and tryptophan.
When separated by SDS-PAGE, prolamins are resolved
into polypeptides of Mr 27, 22, 19, 16, 14 and 10 kDa.
Because of the variation in extraction procedures, the
nomenclature of these proteins is somewhat complex.
The proteins may be divided into groups designated α-,
β-, and γ-zeins, that correspond to the Mr 22,000 and
19,000, the Mr 16,000 and 14,000, and the Mr 27,000
components, respectively. In general, these three
groups are different in their primary amino acid
sequences. The Mr 10,000 protein is structurally
different from the others and may constitute a
distinct fourth group (Shotwell and Larkins, 1988).
Immunogold staining techniques with antibodies
directed against the α- β- and γ-zeins has helped to
better understand the organization and distribution of
the zein within protein bodies isolated from developed
maize endosperm. In general, the α-zeins occur in the
central region of the protein body with the β- and γ-
zeins located primarily at the periphery (Lending et
al., 1988; Wallace et al., 1988a). The location of
the latter two at the protein body surface suggests
that perhaps they are involved in the determination of
its size. It is possible that the synthesis of
certain types of zeins is more prevalent in some
regions of the endosperm than others. Experiments are
in progress (Dr. B. A. Larkins, Dr. A. R. Kriz

laboratories) to resolve the extent to which the process by which protein bodies are assembled affects the endosperm texture (i.e., soft, floury vs hard, vitreous).

Recent findings have shown that introduction of as many as two lysines and two tryptophans into modified zeins and injected into <u>Xenopus</u> oocytes has no major effect on α-zein aggregation, even when inserted into a region thought to be important to zein tertiary structure (Wallace et al., 1988b). This suggests the possibility of eventually expressing high lysine zeins, or perhaps other useful proteins, in the corn endosperm. However, it is unknown what will happen to aggregation of lysine-containing zeins when introduced into corn endosperm.

The genes controlling a number of the α-zeins are on chromosomes 4, 7, and 10. A gene encoding an Mr 15,000 β-zein has been mapped to the long arm of chromosome 6, and the genes encoding the γ-zein have not been mapped. The gene corresponding to the Mr 10,000 zein protein has been mapped to the short arm of chromosome 7. The α-zeins may be organized into multigene families of cDNA clones based on different hybridization criteria. The cDNA clones hybridize to multiple coding sequences in the genome, leading to estimates of between 75 and 150 genes.

Mutant Regulation of Prolamins

Several endosperm mutants that reduce prolamin synthesis have been identified. These mutations are generally recessive. The nutritional quality of the endosperm protein is improved as a consequence of the reduction in prolamin content. In particular, the relative content of the essential amino acid lysine (and tryptophan) is increased, thus the reference to these mutants as "high lysine mutants".

The most important of these mutants and one of the first to be characterized, is opaque-2 (<u>o</u>2) (Mertz, et al., 1964). Introduction of the mutant markedly reduces fraction II (true zein) and increases significantly fraction I (albumins and globulins) and fraction V (true glutelins). These changes have a marked effect on the amino acid composition because

the protein fractions differ widely. Lysine, for
example, is the most variable, varying from 0% in zein
to 6.2% and 6.7% in the albumin-globulin and true
glutelins, respectively. Most zein groups are
affected by opaque-2 mutations, and many alleles
almost completely eliminate the Mr 22,000 α-zeins.
Other non-allelic "high lysine" mutants identified
with similar effects are floury-2 (_fl_2), opaque-7
(_o_7), opaque-6 (_o_6), floury-3 (_fl_3), defective
endosperm-B30 (De[*]B30), mucronate (Mc) and others not
mentioned (see Glover and Mertz, 1987).

No mutants in corn have been reported to alter
substantially the synthesis of glutelins or any
glutelin subunit. Mutants for globulin synthesis are
reported to be expressed only in the germ. However,
recent evidence suggests that globulin genes also may
be expressed in the endosperm of corn. This may be a
genotype dependent effect (personal communication, A.
R. Kriz, Agronomy Dep. Univ. Illinois). Since only a
few genes are involved, this finding may offer some
interesting possibilities for nutritional improvement
by increasing the expression of the natural gene
products by alteration of the promoter regions of
their nutritionally balanced protein.

Protein Quality Improvement

High-lysine Opaque-2 Hybrids. Major emphasis in
most breeding programs for protein quality improvement
is based on the opaque-2 (_o_2) mutant and genetic
modifiers of the endosperm textural properties of
opaque-2 corn (see Glover and Mertz, 1987; National
Research Council, 1988.) As a young faculty member
at Purdue University in 1963, I shared the excitement
of the discovery of opaque-2 corn by E. T. Mertz, O.
E. Nelson, Jr., and L. S. Bates. The implications of
the _o_2 discovery (Mertz, et al, 1964) were considered
remarkable. Later, however, many practical problems
and limitations arose with the standard _o_2 (soft-
endosperm=SE) corn. The failure of _o_2 corn to become
more accepted was primarily due to reduced grain
yield. Kernels of _o_2 are characterized by a soft,
chalky, nontransparent appearance with very little
hard, vitreous (horny) endosperm. This type of kernel

is more prone to damage by insects and kernel rots
both in the field and in storage and harvesting
machinery. Added to these agronomic problems,
beginning in about 1974, were the vacillating opinions
regarding nutritional requirements and priorities
which frustrated the basic requirements for advancing
quality breeding aims, i.e., the issue of calories vs.
protein.

Grain yields of earlier $\underline{o}2$ (high-lysine) hybrids
were 85 to 92% of those of normal dents. The
variation in yield of a U.S. commercial $\underline{o}2$ hybrid from
86 to 92% of its normal counterpart is shown during a
4-year period (Table 1). The data from the 10 tests
indicate the lack of stability in grain yields
associated with year to year variations in different
environments. Definite improvements have been made in
resistance to kernel and ear rots and some
improvements in grain yield of standard $\underline{o}2$ germplasm.
The commercial high-lysine ($\underline{o}2$) cultivars marketed by
Crow's Hybrid Corn Co., for example, have been
improved and their better hybrids yield 89 to 95% of
normal check hybrids. When high-lysine hybrids are
grown in the area to which they are best adapted, they
generally will yield within 5% of normal check hybrids
(C. Malone, personal communication).

TABLE 1

Grain Yields of Commercial Normal Maize Hybrid and Opaque-2 Counterpart grown in Four Different Years.

Year	No. of Tests	Genotype	Yield Mg ha^{-1}	% of normal
1	4	Normal	11.28	
		Opaque-2	10.32	91.4
2	4	Normal	10.82	
		Opaque-2	9.99	91.5
3	1	Normal	9.24	
		Opaque-2	8.22	89.5
4	1	Normal	12.36	
		Opaque-2	10.62	86.0
Overall mean of 10 tests				90.7

Source: Glover and Mertz, 1987.

In 1982, the South African Department of
Agriculture released a commercial yellow high-lysine
hybrid "HL2", a modified single cross (Table 2).

TABLE 2
Performance of South African high-lysine maize

Cultivar	Grain yield	Percent rotton cobs	Percent total lodging	Percent sprouting	Cobs per plant
QPM					
HL2	5.20	5.45	9.05	17.59	1.12
Normal Maize					
PNR 6549	5.56	6.20	9.37	16.57	1.15
RO 405	4.84	6.80	8.15	10.74	0.96
SSM 2045	5.00	7.19	12.62	12.89	1.05
RS 5206	5.13	7.23	12.51	10.92	0.98
PNR 6514	5.26	7.35	6.89	14.27	1.09

Source: South Africa Summer Grain Centre. The normal-
maize figures are for the top five cultivars out of 49
tested in the Centre's national cultivar trials
(1986/87 season).

"HL2" is at a similar yield and agronomic level as the
best normal commercial hybrids. It has hard
endosperm, nearly indistinguishable from normal, and
it has extremely good resistance to ear-rot diseases.
A white high-lysine $\underline{o}2$ hybrid, "HL1", was released
commercially in 1979.

The standard $\underline{o}2$ corn hybrids would appear quite
acceptable for more temperate areas of the world where
corn is used primarily in the feed-grain industry,
provided of course that they were nutritionally
efficient (Bressani, 1976) and/or economically
competitive with the better yielding normal materials.
It does not have good dry-milling characteristics. In
some parts of the world soft-endosperm, floury corns
are preferred such as in the highlands of Bolivia,
Peru, and Ecuador. The appearance of the soft
endosperm $\underline{o}2$ is unacceptable in many tropical and

subtropical areas of the world from the standpoint of
human preference and some associated undesirable
characteristics.

   <u>Modified Opaque-2--Quality-Protein Maize</u>. Much
improvement has been made in increasing resistance to
ear and kernel rots by selection for modifiers of the
<u>o</u>2 gene to enhance kernel vitreousness (hardness).
Modified vitreous endosperm <u>o</u>2 corn is similar to
conventional corn but the kernels are homozygous for
the <u>o</u>2 allele. Researchers have indicated a few genes
to quantitative inheritance and additive genetic
variation for modification of <u>o</u>2 to vitreous hard
endosperm. Reciprocal differences (dosage effect)
between modified by unmodified crosses, and xenia
effects have been observed. In general, there is a
negative relationship between vitreousness or hardness
and protein quality, so selection must be accompanied
by chemical analysis of the endosperm to maintain high
levels of lysine and tryptophan. The protein
fractions of the endosperm portions (soft vs. hard)
are altered differently through selection for hard
endosperm (see Glover and Mertz, 1987). This
observation suggests that the synthesis of certain
types of zeins may be more prevalent in some regions
of the endosperm than others as noted earlier.

   The International Corn and Wheat Improvement
Center (CIMMYT) has had the most extensive and
sustained breeding program in overcoming problems
associated with quality protein improvement in
tropical and subtropical germplasm (Vasal, et al.,
1984, Glover and Mertz, 1987, National Research
Council, 1988). Through backcrossing and recurrent
selection, breeders were able to combine a fairly
complex system of genetic modifiers and the <u>o</u>2 gene to
convert normal maize populations into what is now
referred to as quality protein maize (QPM).

   A conscious effort to maintain protein quality has
been made by the CIMMYT. QPM materials contain about
50% more of the essential amino acids lysine and
tryptophan in the whole grain than does normal.
Almost complete modification to hard endosperm
resembling the normal kernel phenotype has been
accomplished. Selection is still needed to reduce the
variation within an ear although this is becoming less

of a factor. Moisture contents and dry down time are
similar to normal. Ear rot incidence in QPM versions
has been substantially reduced, but ear rots still
affect QPM materials more than their normal versions.
Husk cover in QPM materials also needs improving.

Yield improvement has been observed in some QPM
gene pools but not in others. By and large there has
been continuous improvement in yield performance of
the QPM over the years. Some of the best performing
QPM populations are of tropical adaptation and their
yields compare favorably with corresponding normal
materials (Table 3). Many QPM experimental varieties
evaluated in several countries of Latin America and
the Caribbean have demonstrated yields comparable to
those of normal corn (Table 4). These yields, when
considered along with the nutritional efficiency as
suggested by Bressani (1976), further favor the high-
lysine (QPM) corn.

Despite QPM's promising potential, uncertainties
over its performance in general practice still exist.
Stability of yield over repeated seasons in local
environments, over different locations, and in special
environments with specific disease, insect, or unique
climatic problems is not certain. The question is
aggravated since there are no available near-isogenic

TABLE 3
Performance of Normal Materials and Corresponding
Tropical QPM germplasm, 1987

Material	Grain yield[†]		QPM as percentage of normal
	Normal	QPM	
	-- kg h^{a-1} --		%
Pool 23	5405	5330	98.6
Pool 24	5706	5457	95.6
Tropical High-oil	5733	5170	90.2
Population 62	5347	5484	102.6
Population 65	5255	5369	102.2
Population 63	5705	6236	109.3

Source: Bjarnason and Short, 1988.
[†] Pooled over three locations - Poza Rica and
Tlaltizapan, Mexico and Cuyuta, Guatemala.

TABLE 4
Performance of Quality-Protein Maize (QPM) in Experimental Cultivar Trials, 1986

Entry	Mean Yield	Percentage of normal — Best Check	Percentage of normal — Ref. Entry Check
	Kgha^{-1}	%	%
EVT 15A (17 locations)			
Across 8363	4729	90.1	119.8
Poza Rica 8464	4948	94.5	127.7
San Jeronimo 8464	5124	99.3	131.8
Palmira (1) 8464	4919	91.9	124.9
Across 8464	4980	96.0	126.4
(Range of 20 cv. in test)	(4128-5124)	(78.7-99.3)	(100-131.8)
EVT 15B (21 locations)			
Nioro 8468	5122	92.3	103.9
Nioro (1) 8468	5041	91.2	102.5
Tlaltizapán 8468	5346	96.2	108.9
Tlaltizapán 8369	5018	90.3	102.3
Tlaltizapán 8470	5106	90.3	103.2
(Range of 11 cv. in test)	(4402-5346)	(78.6-96.2)	(95.6-108.8)

Source: Bjarnason and Short (1988).

yield comparisons of QPM materials grown under extensive environments. There may well be some residual heterozygosity in the populations tested and if such were the case, the populations _per se_ would not be expected to perform quite as well on the average. Stability of the kernel modification within the ear and the variation in expression when grown over a number of environments is another question. Acceptability for human consumption and other uses in specific target areas where it would most likely have its greatest impact is an important question. There is a need for cultivar and product customizing for localized situations. In this regard, the CIMMYT's future direction on the QPM program is more narrowly focused than before, emphasizing the development of more elite germplasm for special target countries

where QPM is assessed to have the greatest potential
impact (Bjarnason and Short, 1988). The CIMMYT is
channeling most of its QPM resources into the
identification and improvement of the best performing
QPM population materials for development of inbred
lines from which hybrids or narrow-based synthetics
can be formed. It is assumed, in fact imperative,
that programs for production and utilization of QPM
will require some sort of delivery system or seed
industry to ensure a system of seed production which
would encourage QPM adoption and utilization.

The substantial progress in developing QPM
germplasm with acceptable seed appearance and grain
yield has generated some renewed interest. Guatemala
has released an open-pollinated QPM cultivar (Tuxpeño-
1 H.E.o2) under the local name 'Nutricta'. Nutricta
yielded as much as some of the most popular cultivars
in regional trials (Table 5). QPM hybrids are
presently under development for Guatemala. In 1988,
yield of the best QPM single cross hybrid was 15%
greater than the best normal check hybrid grown in the
lowland area of Guatemala. Experiences with QPM are
beginning to accumulate in various parts of the world.

TABLE 5
Performance of Quality Protein Maize (QPM) cv.
Nutricta in a Central American Regional Variety Trial

Entry Check	Yield	Percentage of
	$Kg\ ha^{-1}$	%
Nutricta (QPM)	5298	98
ICTA-B-1 (Normal)	5235	98
CENTA H-5 (Check)	5397	100
Diamantes 8043	5248	97
Tocumen 7428	5095	94
La Maquina 7727	4780	88
ICTA HB-83	6189	114
CENTA HE 20	6008	111

Data from CIMMYT (Source: Glover and Mertz, 1987).

Some countries have already released QPM materials
(Table 6) although the area planted to them is

limited. Contrary to earlier expectations, the
strongest demand pull could come from animal feed
rather than direct human food uses. Availability and
cost of supplemental proteins for balancing the
rations of monogastric animals will be the key
determinant on the feed side.

TABLE 6
**Use of QPM in National Programs of Developing
Countries**

Country	QPM Germplasm
Guatemala	ICTA-'Nutricta' (released)
Honduras	Nutridia
Bolivia	Tuxpeño 02, Chuquisaca 7741
Ecuador	INIAP 528 (Across 8363)
Paraguay	Nutri-guarini-V-241
El Salvador	Centra M5B
Venezuela	Across 7740, FUNIT 2
Brazil	BR 451
Argentina	RAE-10 (Opaco Semidentado)
Peru	Tuxpeño-1 (QPM), Opaco Huascaran (Comp. J)
Senegal	Obregon 7740, Temperate White QPM
China	Tuxpeño 102, Hybrid 203 (SC 201 x line 079)
Mexico	Composite 1, Tuxpeño-1 QPM, Poza Rica 8140, Guanacaste 7940 Population 68
Haiti	Population 65
Costa Rica, Honduras, Venezuela, Panama Dominican Rep., Argentina, Guatemala, Peru, Brazil, China	Various QPM pools and populations for derivation of inbred lines

Source: Glover and Mertz (1987).

In the U.S., the Texas A & M University group is
actively researching CIMMYTs' QPM materials,
extracting early generation inbreds and producing some
experimental hybrids adapted to the Southern U.S. At
Purdue University, we have continued a modest program
to develop a number of elite Corn Belt inbred lines

with the hard endosperm _o_2 (QPM) characteristics. We
are also researching four of CIMMYT's temperate QPM
populations that are quite well adapted to our Corn
Belt conditions. The University of Illinois has
continued with a small program of developing hard
endosperm _o_2 germplasm adapted to the Corn Belt.
    Sugary-2;Opaque-2 Hybrids. At Purdue University,
we have also investigated the use of interactions
between genes regulating storage proteins and genes
influencing starch characterisitcs to improve protein
quality in corn. The double mutant between _o_2 and the
starch-modified gene sugary-2 (_su_2) has been
investigated most extensively (Glover and Mertz,
1987). The unique features of the _su_2;_o_2 kernel
phenotype are the characteristic translucent, hard,
vitreous endosperm texture and kernel density that
compares favorably with the normal counterpart dent
hybrids (Table 7). The _su_2 and _su_2;_o_2 starches are

TABLE 7
Grain, Protein, and Lysine Yields; and Kernel
Characteristics from 25 near-isogenic _opaque_-2,
sugary-2, and _sugary_-2; _opaque_-2 Single Crosses and
Their Normal Counterpart Hybrids. Mean values from
Replicated Trials

| Character | Genotype | | | |
	Normal	_su_2	_o_2	_su_2;_o_2
Grain yield (mg ha^{-1})	9.80	8.59	8.38	7.85
Protein (kg ha^{-1})	921.0	880.3	800.3	802.6
Lysine (kg ha^{-1})	26.1	28.0	34.7	38.9
Kernel density (g cm^{-3})	1.27	1.33	1.10	1.24
100-Kernel wt. (g)	32.1	28.2	25.9	24.5
100-Kernel volume (cm^3)	25.2	21.2	23.5	19.8
Grain test-wt. (kg m^{-3})	778.9	802.8	643.0	699.0
Wisconsin breakage test (%)	4.01	1.53	5.00	1.33
Missouri breakage test (%)	25.8	17.1	33.7	19.1

Source: Glover and Mertz (1987).

more digestible than normal starches as indicated by
relative rates of animal and bacterial amylase enzyme

hydrolysis of starches. This characteristic may
contribute toward improved biological value (Glover
and Mertz, 1987). The su2;o2 kernels resist kernel
breakage and mechanical stress better than o2 kernels.
Test weights of su2;o2 hybrids are significantly
better than o2, but they are inferior to normal
versions.

The su2;o2 modified corn reduced kernel size
(volume) and dry weight, and produced grain yields
that were 80% of the normal hybrids. Protein yield
was 87% of the normal hybrids. In spite of the lower
yields, the su2;o2 hybrids averaged 12% more lysine
ha^{-1} than the o2 hybrids and 49% more lysine ha^{-1} than
the normal hybrids.

An alternative approach to improving protein
quality in corn has been by tissue culture and in
vitro methods involving amino acids or amino acid
analogues. To date, no new cultivars have been
developed from selected mutants for release to the
grain producer (Gengenbach, 1984; Hibberd, et al.,
1986).

Nutritional Value of Quality Protein Corn

The nutritional superiority of high-lysine corn
has been repeatedly demonstrated in trials with
infants, small children, and adults, and in a variety
of animal feeding trials. Recent experiences in
Nebraska trials (Asche, et al., 1985) with high-lysine
hybrids and in South Africa (H. Gevers, personal
communication) with "HL2" demonstrated a savings in
reduced protein level (2% lower) of the high-lysine
corn diets when substituted for ordinary corn in an
animal ration for pigs. The nutritional advantage of
QPM diets for starter and grower pigs has been
demonstrated (Sullivan, 1988). Sugary-2;opaque-2 corn
is similar to o2 in protein quality; and energy
utilization in su2;o2 is similar to normal corn and
more highly digested than o2 corn for growing pigs
(see Glover and Mertz, 1987). The South African
researchers have indicated that high-lysine "HL2" may
be of significant value in the diet of ruminants,
i.e., dairy cattle. There are testimonials of U. S.
farmers who feed o2 corn silage to dairy cattle that

claim benefits in increased milk production of their
better producers. QPM surpasses normal corn in
biological value and net protein utilization (Table
8), and although highly digestible, the true
digestibility is slightly inferior to normal corn. It
now appears that the quality and biological value of
QPM protein is about 90 percent of that of

TABLE 8
**Nutritional Evaluation in Rats of Normal, $\underline{o}2$,
and QPM Corn**

| | | Response criteria | | |
Sample	Lysine in protein	True digesti- bility	Biolo- gical value	Net protein utili- zation
	%	%	%	%
Blanco Cristalino (normal)	2.6	98.1	62.7	61.5
Tuxepño $\underline{o}2$ (Soft E)	4.2	96.0	77.6	74.5
Amarillo Dentado (QPM)	4.1	95.8	74.5	71.4
Ant. X Ver. 181 (QPM)	4.0	96.6	76.2	73.6

Source: Vasal et al. (National Research Council,
1988).

milk protein vs around 40 percent for common corn
protein. QPM was almost 50 percent more effective
than common corn at fostering growing in recovering
malnourished children (National Research Council,
1988; G. Graham, personal communication). Dr.
Graham's group have recently demonstrated that QPM, as
the only source of protein and fat in the diet of
weaned infants and small children, can support growth
equivalent to that attained with cow's milk-derived
formulas. Whole kernel $\underline{su2};\underline{o}2$ corn diets also
approached adequacy in meeting the energy and protein

needs of recovering malnourished infants and young
children (Graham et al., 1980).

QPM shows promise for use in commercial food
products. A 20 percent whole QPM flour can be blended
with wheat flour without affecting the loaf value and
crumb quality of leavened breads (Protein Quality
Laboratory, CIMMYT). A QPM flour blend is presently
being utilized in Guatemala to make nutritionally
fortified cookies (Bressani, personal communication).
QPM gives a good yield of flaking grits and has
outstanding promise in unleavened foods such as
tortillas, corn chips, and other corn-based snack
foods (Sproule et al., 1988). Milling quality as
measured by milling equivalent factor (MEF) of $\underline{su}2;\underline{o}2$
corn shows promise for utilization via dry milling.
It is possible that a high quality food concentrate
could be made from QPM and $\underline{su}2;\underline{o}2$ modified protein
corns.

A recent reassessment of adult requirements for
essential amino acids suggests that the required
levels of lysine, valine, and threonine are likely to
be about two to three times higher than current
requirement figures proposed in 1985 by FAO/WHO/UNU
(V. Young et al. 1987, personal communication).
Using these new estimates, they found lysine is
limiting in the diets of a number of countries, and
recommended therefore that the quality and quantity of
dietary protein continue to be an important feature of
the design and implementation of food nutrition and
agricultural programs and policies.

The potential advantages of quality protein corn
(QPM) are enormous, and it will probably have a role
to play in human nutritional intervention. But the
exact nature of that role remains uncertain.
Certainly the potential role of improved protein
quality corns in animal feeding could carry an even
greater impact, particularly in view of the upward
trends in the global use of corn as livestock feed and
the indirect benefits to humans.

References

Asche, G. L., A. J. Lewis, E. R. Peo, Jr., and J. D.
Crenshaw. 1985. The nutritional value of normal
and high lysine corn for weaning and growing-
finishing swine when fed at four lysine levels.
J. Anim. Sci. 60:1412-1428.

Bjarnason, M., and K. Short. 1988. The current status
and future strategies of the quality protein
maize program. CIMMYT, Mexico, DF.

Bressani, R. 1976. Productivity and improved
nutritional value in basic crops. p. 265-287. In:
H. L. Wilcke (ed.) Improving the nutritional
quality of cereals II. Life and Agency for
International Development, Washington, D.C.

Gengenbach, B. 1984. Tissue culture and related
approaches for grain quality improvement. p. 211-
254. In: G. B. Collins and J. G. Petolino (eds.)
Applications of genetic engineering to crop
improvement. Martinus Nijoff/Dr. W. Junk Publ.,
The Netherlands.

Glover, D. V., and E. T. Mertz. 1987. Corn. Chapter 7.
In: R. A. Olson and K. J. Frey (eds.) Nutritional
Quality of Cereal Grains: Genetic and Agronomic
Improvement. American Society of Agronomy,
Madison, WI., Agronomy Monograph 28:183-336.

Graham, G. G., D. V. Glover, G. L. de Romaña, E.
Morales, and W. C. MacLean, Jr. 1980. Nutritional
value of normal, opaque-2 and sugary-2 opaque-2
maize hybrids for infants and children. I.
Digestibility and utilization. J. Nutr. 110:1061-
1069.

Hibberd, K. A., M. Barker, P. C. Anderson, and L.
Linder. 1986. Selections for high tryptophan
maize. p. 440. In: D. A. Somers et al. (eds.)
Abstr. VI Int. Congress of Plant Tissue and Cell
Culture. Univ. of Minnesota, Minneapolis, MN,
August 3-8.

Kauffmann, K. D., and J. W. Dudley. 1979. Selection
indices for corn grain yield, percent protein and
kernel weight. Crop Sci. 19:583-588.

Lending, C. R., A. L. Kriz, B. A. Larkins, and C. E. Bracker. 1988. Structure of maize protein bodies and immunocytochemical localization of zeins. Protoplasma 143:51-62.

Mertz, E. T., L. S. Bates, and O. E. Nelson. 1964. Mutant gene that changes protein composition and increases lysine content of maize endosperm. Science 145:279-280.

National Research Council. 1988. Quality-Protein Maize. National Academy Press, Washington, D.C. 100 p.

Shotwell, M.A., and B. A. Larkins. 1989. The biochemistry and molecular biology of seed storage proteins. In: Biochemistry of Plants: A comprehensive treatis: Vol. 15:297-345.

Sproule, A. M., S. O. Saldivar, A. G. Bockholt, L. W. Rooney, and D. A. Knabe. 1988. Nutritional evaluation of tortillas and tortillo chips from quality protein maize. Cereal Foods World 33:233-236.

Sullivan, J. S. 1988. Nutritional value of quality-protein maize food and feed corn for starter and grower pigs. M. S. Thesis. Texas A & M Univ., College Sta., TX. 42p.

Vasal, S. K., E. Villegas, and C. Y. Tang. 1984. Recent advances in the development of quality protein maize germplasm at CIMMYT. p. 167-189. In: Panel proceedings series: Cereal grain-protein improvement. 6-10 Dec. 1982. IAEA, Vienna, Austria.

Wallace, J. C., G. Galili, E. E. Kawata, C. R. Lending, A. L. Kriz, C. E. Bracker, and B. A. Larkins. 1988a. Location and interaction of the different types of zeins in protein bodies. Biochem. Physiol. Pflanzen. 183-107-115.

Wallace, J. C., G. Galili, E. E. Kawata, R. E. Cuellar, M. A. Shotwell, and B. A. Larkins. 1988b. Aggregation of lysine-containing zeins into protein bodies in Xenopus oocytes.

QUALITY PROTEIN MAIZE - WHAT IS IT AND HOW WAS IT DEVELOPED*

E. Villegas
Former Head of CIMMYT's Service Laboratory
Sur 71B No. 156 Bis, Col. El Prado
Delegation Iztapalapa
09480 Mexico, D.F.

S.K. Vasal, and M. Bjarnason
CIMMYT Maize Breeders
CIMMYT, Lisboa 27, Apdo Postal
6-641, Mexico 6, D.F.

Common or normal maize grain is generally low in protein content (average 9.5%) and moreover, deficient in protein quality, as a consequence, mainly, of the low levels of lysine and tryptophan (1.70 and 0.40%) in the endosperm protein. Both are essential amino acids that humans and monogastric animals, such as swine and poultry, must obtain from food because they cannot synthesize them. Roughly, half of the protein in a kernel of normal-maize is composed of protein fractions that contain almost no lysine and tryptophan, the latter an essential amino acid that is also a precursor of the important B-vitamin, niacin.

---

*   Presented at the Annual Meeting of the American Association of Cereal Chemists, Dallas, Texas, October 17, 1990.

<u>Discovery of high lysine maize</u>. Since the early part
of the 20th century, scientists have sought to breed a
more nutritive maize and the research was aimed at
enhancing the protein concentration of the grain. Two
general approaches have been used to increase the
protein content of cereal grains, namely: 1)
application of fertilizers, and 2) breeding improved
varieties. The protein content of maize can be
modified by selection and breeding. The classical
selection experiment for high and low protein content
in the grain conducted at the Illinois Agricultural
Experiment Station, provided valuable insights into
the possibilities and limitations of recurrent
selection for single polygenic traits. After 70
generations of selection, protein content had
increased from 10.9% in the original population to
26.6% in the Illinois High Protein strain (Dudley,
1974). High and low-protein lines of maize were
analyzed for lysine (Miller et al. 1950) and
tryptophan (Frey 1951) and results showed that the
content of these amino acids in the protein decreased
as the content of the protein in the grain increased.
Increases in the crude protein content of the maize
grain, either due to fertilization or selection,
increased the proportion of zein (Jimenez 1966), that
contains almost no lysine and tryptophan.

For many years, plant scientists believed that
the shortcomings of cereal protein quality, i.e.
inadequate amino acid balance, could not be corrected
by genetic biochemical manipulation. Then came the
discoveries at Purdue University in the early 1960s
(Mertz et al. 1964; Nelson et al. 1965) that certain
mutant genes in maize could greatly improve the amino
acid composition of the endosperm protein. This
discovery opened new doors to plant breeders in the
improvement of the nutritive value of cereal proteins.

In 1968, Swedish scientists reported the first
positive variation in the amino acid composition in a
primitive Ethiopian barley variety from the World
Barley Collection (Munck et al. 1969), and three years
later another induced mutant gene that increased
considerably the level of lysine in the barley protein
was described by Ingversen. In 1973, the first high

lysine sorghum mutant was reported at Purdue
University (Singh and Axtell 1973).

<u>Plant breeding studies with opaque-2 maize</u>. After the
discovery of the biochemical effects of the opaque-2
gene (Mertz et al. 1964) and other mutant genes such
as floury-2 (Nelson et al. 1965), opaque-7 (McWhirter
1971) and others that modified the protein quality of
the maize grain by increasing the level of lysine and
tryptophan in the endosperm protein, great enthusiasm
was generated worldwide among maize breeders, and in
maize breeding programs in many countries. CIMMYT
initiated the conversion of normal maize to opaque-2
or floury-2 materials mainly by backcrossing although
other approaches were also utilized in order to obtain
high lysine maize as rapidly as possible. These
efforts led to the development of an array of opaque-2
varieties and hybrids in the late 1960s and early
1970s, some of which were grown commercially in
Brazil, Colombia, United States, India, Yugoslavia,
Hungary and South Africa. In most instances, grain
yields were disappointing. The high lysine mutants
adversely affect several agronomic characteristics, as
well as the kernel phenotype. Well known drawbacks of
early opaque-2 maize cultivars have been:
a) Reduced accumulation of dry matter, lower weight
    and density of the kernel which is reflected in
    lower yield than in their normal counterparts.
b) Texture of the kernel, which with o2 gene becomes
    soft, chalky and dull and this type of kernel is
    not very well accepted by consumers in many
    countries.
c) Greater susceptibility to ear rots.
d) More susceptibility to grain pests, especially in
    storage.
Studies initiated in some countries to examine the
problems in the production and acceptability by
consumers of opaque-2 maize in farmers' fields,
confirmed the seriousness of the problems of the
converted opaque-2 materials (Pinstrup-Andersen 1971;
Francis et al. 1972; Sreeramulu and Bauman 1970;
Nelson 1966; Harpstead 1969; Lambert et al. 1969; Paez
et al. 1969; Arriaga et al. 1970; Gevers 1972; Singh

and Asnani 1975; Vasal 1975; Vasal et al. 1980, 1984a, 1986).

Not all of the problems were of equal relevance in all parts of the world. Reduced yield, for example, the single most important limitation, weighted most heavily in the United States and other developed countries where average yields of normal hybrids were higher. In developing countries, where farmers are accustomed to growing hard flints and dents, the soft opaques had greater susceptibility to ear rots and stored grain pests, and thus were not well accepted by farmers and consumers. In addition, it was reported that some genotypes of opaque-2 maize sown at low temperatures were adversely affected in plant emergence. Also, the soft kernels of opaque-2 maize present problems for machine harvesting (damaged, broken or cracked kernels), and for the milling industry.

<u>Breeding for improved kernel quality in opaque-2 maize germplasm</u>. In an effort to reduce the defects of opaque-2 maize, many scientists in different countries experimented with various approaches in high protein quality germplasm development. Among the most promising methods were:
a) Recurrent selection for high lysine content
b) Increase in germ size
c) Development of double mutant combinations with
    normal kernel phenotype
d) Development of hard endosperm by the use and
    accumulation of modifier genes for the locus o_2.
In each case, the principal aim was to develop quality protein maize with normal looking appearance and high yield (Paez et al. 1969; Paez and Zuber 1973; Paez et al. 1975; Vasal 1975; Misra 1975; Zuber 1975; Glover et al. 1975; Bjarnason et al. 1976; Vasal et al. 1980).

<u>Recurrent selection</u>. There were several problems in developing high lysine maize through recurrent selection in normal endosperm maize populations, mainly the small variability in the level of the amino acids, lysine and tryptophan, the lack of simple

chemical methods to detect this variability and the
difficulty of transferring the high lysine trait to
other backgrounds (Paez et al. 1969; Vasal et al.
1980, 1984a).

Double mutants. In the approach of double mutants,
the floury-2, opaque-2 and sugary-2, opaque-2
combinations were investigated as a means of
developing normal looking protein quality maize. The
floury-2, opaque-2 combination in some backgrounds
generated vitreous kernels (Nelson 1966), however,
several of these combinations were not necessarily
vitreous (Vasal et al. 1980). Because the interaction
between two mutants occurs only in rare backgrounds,
this approach was discarded at CIMMYT and was
exploited with only limited success in other
institutions. The combination sugary-2, opaque-2
showed several advantages and has been studied by some
scientists (Paez 1973; Glover et al. 1975; Vasal et
al. 1980 and Glover 1988). The intereaction between
the two mutant genes in a homozygous recessive
condition results in vitreous kernels whose lysine
content in the endosperm protein is sometimes higher
than that of opaque-2 maize. However, the kernels are
smaller and weigh 15-25% less than the normal ones,
depending upon the genetic background (Vasal 1984).
This combination appears promising for improved kernel
phenotype but has very poor grain yield. At CIMMYT a
sugary-2, opaque-2 composite was developed by pooling
families that exhibited good phenotype and differed in
kernel weight compared to soft opaque-2. Cycle 4
showed reduced days to flowering, ear height, and
moisture content of the grain at maturity, compared
with Cycle 0 but later, better results were obtained
with other approaches.

Use of modifier genes. The third approach employed at
CIMMYT was to develop quality protein maize germplasm
through the combined use of two genetic systems: the
opaque-2 gene and the genetic modifiers which
contribute to kernel modification (hard endosporm) and
maintain acceptable levels of the protein quality.
Initial efforts began as early as 1970 when the

breeders began selecting partially modified kernels
from opaque-2 composites and the backcross-derived
opaque-2 populations (CIMMYT 1970).

<u>Analysis for protein quality</u>. A protein quality
laboratory was established at CIMMYT in the late 1960s
to carry out chemical analyses on thousands of small
samples from individual ears of the many QPM breeding
populations and even individual seeds from segregating
ears. To handle this large volume of samples, the
cereal chemists developed efficient methods for
determining protein quality (Hernández and Bates 1969;
Tsai et al. 1972; Paulis et al. 1974; Mossberg 1969;
Villegas and Mertz 1970; Mertz 1974; and Villegas et
al. 1984). Most of these methods have been adopted in
various protein quality screening laboratories around
the world. Many young scientists from developing
countries were trained at CIMMYT in laboratory
techniques using simple techniques and inexpensive
equipment for determining protein content, tryptophan
and/or lysine content, and/or zein content.

From studies of the relationship between kernel
modification and protein quality in various classes of
kernels, it was obvious that selection for kernel
modification could adversely alter quality protein
traits; however, variation existed in the content of
tryptophan and lysine in the endosperm protein, and
the selection for hardness could be performed without
sacrificing the protein quality (Table 1).

The strong support from the CIMMYT chemical
laboratory permitted progress to be made in the
selection of modifier genes for endosperm hardness and
improved protein quality. Experience in breeding QPM
at CIMMYT indicates that during recurrent selection
for kernel modification, the protein quality of
individual progenies must be carefully monitored to
maintain it at a high level (Vasal 1975; Vasal et al.
1975; Ortega and Bates 1983; Bjarnason et al. 1988);
some modifiers for endosperm hardness reduce protein
quality (Table 2).

Table 1. Lysine and tryptophan content in endosperm
          of normal, opaque-2 soft, and opaque-2
          modified maize versions.

	Normal	Opaque-2 Soft	Opaque-2 Modified
Lysine % in Protein	2.00	3.80	3.40
Tryptophan % in Protein	0.40	1.00	0.85
Protein % (% N x 6.25)	9.20	8.00	9.93

Table 2. Protein, lysine and tryptophan content in
          endosperm of three QPM populations through
          different cycles of selection.

QPM Populations	Cycle	% Protein	% Tryptophan in Protein	% Lysine in Protein
Yellow QPM	C_0	8.0	0.78	2.90
	C_6	8.0	0.75	2.90
PD (MS) 6 QPM	C_0	8.5	0.78	2.50
	C_4	8.8	0.78	2.60
Temperate x Tropical QPM	C_0	8.0	0.66	3.20
	C_8	8.0	0.68	2.90

The research on hard endosperm opaque-2 maize
passed through various exploratory stages. A major
effort was initiated to select partially modified ears
from as many different opaque-2 conversions as
possible in tropical, subtropical and highland QPM
germplasm. Initially, chemical analyses for
tryptophan were carried out on microsamples extracted
from tiny bands or spots of hard endosperm using a

small electrical drill. Since this procedure did not
damage the embryo, kernels containing the protein
quality desired were replanted. As the kernels became
more modified, the chemical analyses for protein and
tryptophan content were performed on the whole
endosperms from 10 kernels per ear.

<u>Population improvement</u>. With the support of national
programs in developing countries, yield trials were
conducted globally and in Mexico with the CIMMYT
opaque germplasm. About 40% of the superior families
were analyzed and selected each year. In the
subsequent growing seasons, reciprocal crosses were
made between individual plants among these families.
At harvest, 250 pairs of ears were analyzed and
selected for the next cycle of evaluation. After each
cycle, kernels from hard-endosperm ears were selected
from the superior families and evaluated for protein
quality. The laboratory analyzed approximately 25,000
samples each year.

<u>Breeding strategies</u>. Modified endosperm opaque-2
donor stocks were developed through two strategies:
a) Development of white and yellow broad based
    opaque-2 composites and subsequent selection for
    modified types with good protein quality.
b) Intrapopulation selection for modified kernels in
    narrow based populations which had enough
    variability for kernel modification.
The donor stocks were later used extensively in
converting a whole range of tropical and subtropical
maize germplasm at CIMMYT to hard-endosperm opaque-2
types through a backcross-cum-recurrent selection
procedure (Vasal et al. 1980). Both in population
improvement and in the conversion program, a multi-
trait selection procedure using independent culling
levels was employed to accumulate modifiers, maintain
improved protein quality, increase resistance to ear
rot, and improve yield and other agronomic traits of
opaque-2 germplasm (Vasal et al. 1986). By the late
1970s and early 1980s, the QPM versions of most of
CIMMYT's tropical and subtropical pools and
populations were developed and when compared in yield

trials with their normal counterparts, gave acceptable
yields as shown in Table 3 (Vasal et al. 1984a). At
this point, the total volume of QPM germplasm was
reduced by merging genetically similar materials which
were then further improved in homozygous opaque-2
backgrounds. Thirteen pools and ten populations
resulting from this process were described by Vasal et
al. (1984a) and Glover and Mertz (1987). The
improvement of QPM pools and populations continued
until 1987. In addition, subtropical yellow and
tropical white high-oil QPM populations were
developed.

Table 3. Comparison of normal materials with their
          QPM versions.

Material	Grain yield[a] T/Ha normal	QPM	QPM as % of normal
TUXPEÑO-1	6.19	6.15	99.4
MIX. 1 COL.			
GPO. 1 x ETO	6.12	5.68	92.9
MEZCLA AMARILLA	5.43	5.28	96.4
AMARILLO DENTADO	5.35	5.23	97.7
TUXPEÑO CARIBE	6.39	5.90	92.3
ANT. REP.			
DOMINICANA	5.25	5.08	94.9
LA POSTA	6.47	5.90	91.2

a From three environments.

Performance of QPM varieties. At present, CIMMYT's
QPM program has at its disposal the collection of
tropical and subtropical QPM pools and populations
which have undergone recurrent selection for endosperm
hardness, several agronomic characters and yield, as
well as experimental varieties formed from the best
full sib families identified through international
miltilocational testing. Today, within the range of
experimental error, certain QPM genotypes equal the
yields of the best conventional maize varieties now
under cultivation in many developing countries where

maize constitutes the main bulk of the daily diet, for
both adults and young children, particularly for the
low-income groups. During the growing seasons of
1983, 1984 and 1985, several experimental QPM
varieties performed better than the normal-maize
checks in several regions of the world. In some
experiment stations, yields have been as high as 6,900
kg/ha. Indeed, three open-pollinated varieties (La
Posta QPM, Obregon 7740 QPM and Tuxpeño-1 QPM) gave
yields close to those of two hybrids and one open-
pollinated variety that have been widely used
commercially. In trials at many locations around the
world, several QPM materials have shown yields equal
or higher than the best of the checks tested (Table
4).

<u>Current program at CIMMYT</u>. In 1985, a small QPM
hybrid program was initiated. In 1987, recurrent
selection in the QPM pools and populations was
terminated at CIMMYT. Currently the QPM program
concentrates its efforts on the development of inbred
lines with good combining ability, good agronomic
traits and disease and insect resistance which can be
used either to develop various types of hybrids, or to
form synthetics for use as open-pollinated varieties.
This change in breeding methodology was made for
several reasons:
a)   Since hybrids permit maximum exploitation of
     heterosis, one can expect better performance from
     hybrids than from open-pollinated varieties
b)   Proper production of QPM hybrid seed would ensure
     its purity, which may be a problem at the farm
     level with open-pollinated varieties in less
     developed countries
c)   Hybrids should be more unfirm and stable in kernel
     modification than open-pollinated varieties, and
d)   Hybrids would require a minimum of protein quality
     monitoring as long as the parent lines are kept
     pure. This is an important consideration, since
     many breeding programs in developing countries do
     not have the support of a protein quality
     laboratory.

Table 4.   Performance of QPM in selected countries (QPMT-11A, 1981).

Country/location	Best check (normal)	Yield (T/ha)	Best QPM entry	% Yield of best normal check
Costa Rico-Alajuela	Tico H4	4.5	Across 7740 Re	104
Guatemala-Jutiapa	Local	4.6	Guanacaste 7940	136
Guatemala-La Maquina	HB-47 (H)	4.6	White Bu Pool QPM	109
Honduras-La Ceiba	Check 1	3.8	Amarillo Dentado QPM	100
Nicaragua-El Plantel	La Maquina 7422	6.1	Ferke (1) 7940	96
Panama-Guarare	Pioneer X-306B	4.4	White Bu Pool QPM	91
Panama-Rio Hato	Tocumen Planta Baja	4.4	Tuxpeño-1 QPM	115
Panama-Tocumen	Tocumen 7428	5.3	Guanacaste 7940	92
Mexico-Obregon	Poza Rica 7822	4.8	Pool 23 QPM	88
Mexico-Obregon (2)	Poza Rico 7822	4.5	Across 7740 Re	114
Mexico-Poza Rica	Population 49 RSF	6.7	Guanacaste 7940	92
Mexico-Nayarit	Criollo Car. Puerto	3.7	White Bu Pool QPM	118
Ethiopia-Nazareth	Shaje	3.7	Across 7839	118
Upper Volta-Farako-Ba	Irat-100	3.5	Across 7740	109
Senegal-Njoro	HVB-1	3.6	Ferke (1) 7940	109
India-Delhi	EH 400175	6.2	Amarillo Dentado QPM	88

CIMMYT's QPM program is developing relatively
homozygous inbred lines and forming three-way and
double-cross hybrids for testing in cooperation with
national programs in different parts of the world.

Additional products of the program are topcross
hybrids, variety hybrids, and synthetics for use as
open-pollinated varieties. The flow of germplasm is
shown in Fig. 1. Elite germplasm from CIMMYT's QPM
program is the main source material used for
developing hybrids, although other sources of QPM
germplasm are employed to a limited extent.

During early generations of inbreeding, the lines
are evaluated for standability and anthesis-silking
interval; they are also evaluated for disease
resistance, as well as for protein quality by chemical
analysis. Testcrosses for evaluation of combining
ability are normally started in the S_3 generation and
not earlier. As the per se performance of the lines
improves, testing for general combining ability at an
earlier stage will be considered. Initially, open-
pollinated experimental varieties were used as
testers, but now single crosses, which can serve as
parents in three-way and double cross hybrids, are
also being used. Diallel crosses are made among lines
with good per se performance and good performance in
testcrosses.

Production of hybrids. Initial results indicate that
very good hybrids can be found among and within some
of the populations. As indicated in Fig. 1, elite
lines are recycled by forming narrow-based synthetics
within heterotic groupings or by crossing two lines
and deriving new lines from this cross. The narrow-
based synthetics could also be used as parents of
variety hybrids (Bjarnason 1990).

It must be pointed out here that many developing
countries still lack the seed production and marketing
capacity necessary to take advantage of hybrids, and
in some cases, the level of agricultural development
is such that hybrids do not show significant advantage
over open-pollinated varieties in farmers' fields.
More appropriate QPM products for these countries, at
least at present, are experimental varieties, which

Figure 1.   Flow of germplasm in the CIMMYT QPM
            Program.

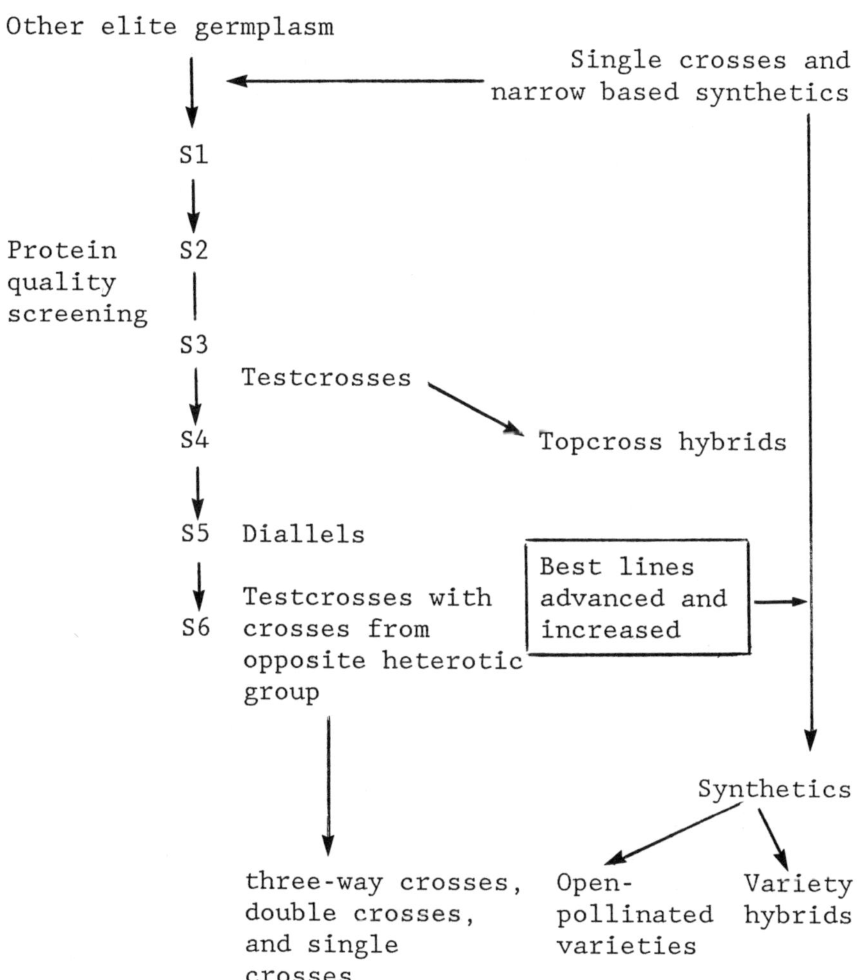

were developed through international testing of QPM
populations and have performed well in many parts of
the world, and synthetics, which are being formed from
elite lines across heterotic groups in the hybrid
program. Advanced generations of these synthetics can
be used as open-pollinated varieties.

Cooperating countries will continue to evaluate
germplasm developed at CIMMYT, both open-pollinated
varieties and hybrids, and identify the best germplasm
for further testing and possible refinement under
their own conditions. Central and South American
countries are already evaluating three-way and double-
cross hybrids, for instance, Guatemala and Brazil. A
yellow QPM double cross hybrid will be released in
Brazil in the next two years for feed use (R.
Magnavaca pers. com.). Also, in many African
countries QPM hybrid germplasm is being evaluated, but
the future course of this work depends on the
development of seed industries.

The QPM breeding program tests and selects
materials for the major diseases and environmental
stresses of the tropics and subtropics in different
sites in Mexico. One exception is the maize streak
virus (MSV) found only in Sub-Saharan Africa and
neighboring islands where it causes serious yield
losses and it is difficult to control except with
resistant varieties. To overcome this handicap,
CIMMYT and IITA (International Institute of Tropical
Agriculture) initiated in Africa a joint program in
1980 for converting CIMMYT's maize germplasm to maize
streak virus resistant, and recently this activity is
continued by CIMMYT's research group at Harare,
Zimbabwe. Also, an effort to develop downy mildew
(*Perenosclerospora spp*) resistant QPM germplasm has
recently been initiated by CIMMYT's regional program
in Thailand.

<u>Yields in developing countries</u>. In Table 5, the yield
results obtained with white late tropical QPM
materials in five countries are shown. The materials
tested were: 3 double crosses, 3 three-way crosses, 1
variety hybrid, 2 open-pollinated varieties compared
with 2 normal local checks. One QPM hybrid yielded

more than the best local check in each place, and the
three QPM hybrids of higher yield, produced
significantly more grain than the best local check
across the five locations. Similar results were
obtained with tropical yellow QPM hybrids in other
trials (Table 6).

Table 5. Results of a tropical QPM trial conducted at
          five locations in 1989.

Entry	Grain Yield	No. of days[1] to silk	Ear[2] height	Ear rot
	T/ha	No.	cm	%
Double cross 2 Q	7.95	58	127	5.6
Three-way cross 2 Q	7.74	60	114	2.1
Double cross 1 Q	7.72	66	121	3.5
Three-way cross 3 Q	7.67	64	134	6.9
Family hybrid Q	7.53	59	125	6.3
Three-way cross 1 Q	7.44	61	138	4.4
Capinopolis 8563	7.07	61	123	8.8
Double cross 3 Q	6.75	62	129	9.2
Las Acacias 8562	6.36	65	110	6.1
Best local check	7.26	71	135	3.9
LSD 5%	0.58			

1 Data from three locations.
2 Date from four locations.
Locations: Mexico: Obregón, Los Mochis; Colombia:
Palmira (1), Palmira (2); Honduras: Danli.

Breeding for improved QPM in other countries. A
limited amount of QPM research is underway in the USA,
as in Purdue University, the University of Illinois,
and Texas A&M University. In the private sector, two
or three seed companies are conducting research on the
development of soft and hard endosperm opaque-2
hybrids (Crow's Hybrid Co., Wilson Hybrids and
recently Pioneer Co.). Interesting work is being
conducted in some other temperate zone countries,
especially in Italy, West Germany, a few East European
nations and the Soviet Union. South Africa has also

had an active hybrid program, which reports some opaque-2 hybrids with very little yield differential compared with normal maize hybrids. It is expected that QPM research may be influenced by further use of tissue culture and future development in molecular biology. A line with elevated methionine concentration (BSSS 53) was identified by screening seedlings on lysine-plus-threonine supplemented media (Phillips et al. 1981). This high methionine maize derived from such *in vitro* methods has not yet been tested for agronomic performance and usefulness in breeding programs.

Table 6. Results from a tropical yellow QPM trial conducted at five locations in 1989.

Entry	Grain Yield	No. of days[1] to silk	Ear height	Ear rot
	T/ha	No.	cm	%
Three-way cross 1 Q	5.54	49	113	8.5
Double cross 2 Q	5.47	53	111	10.5
Double cross 1 Q	5.36	51	116	7.2
Double cross 3 Q	5.26	50	114	11.2
Three-way cross 3 Q	5.23	53	116	6.5
Three-way cross 2 Q	5.14	51	108	6.6
Across 8565	4.57	52	102	12.7
S8766 Q	4.39	51	99	8.0
Poza Rico 8666	4.35	52	104	10.1
Best local check	5.46	56	121	9.2
LSD 5%	0.32			

1   Data from four locations
Locations: Colombia: Palmira; El Salvador: San Andres, Santa Cruz; Honduras: Omonita; Mexico: Obregon.

Progress made in recent years in the development of a restriction fragment length polymorphism (RFLP) linkage map of maize, offers interesting opportunities for accelerating progress in QPM breeding (Schmidt et al. 1987 and Tanksley et al. 1989).

QPM PROGRAMS IN TROPICAL AND SUBTROPICAL COUNTRIES

<u>Use of CIMMYT germplasm in tropical and subtropical regions</u>. Many countries conduct experimental variety trials on QPM from CIMMYT and request seed of the best varieties for increase and further testing. In Central America there is interest in QPM for both food and feed. Guatemala has released an open-pollinated QPM variety named Nutricta that is based on Tuxpeño QPM. Further selection for improved husk cover and adaptation has been conducted locally. National programs in Guatemala and other Central American countries are evaluating white and yellow QPM lines and hybrids from CIMMYT and also are interested in yellow maize for feed use.

In South America, Brazil released in 1988 the open-pollinated variety BR451, which is derived from CIMMYT's population 64 (Blanco Dentado-2 QPM). It has shown good adaptation to many Brazilian environments and is very suitable for maize-wheat flour mixtures for the baking industry. In 1990, 1,410 tons of seed will be available to farmers, enough to plant 70,500 ha (Magnavaca, pers. com.). In Bolivia, soft-endosperm Tuxpeño-O_2 is being grown on about 5,000 ha. In Ecuador, the variety Across 8363, recently released as INIAP-528, has been found particularly suitable for choclos (boiled green ears) or roasted maize. In Venezuela, the variety Across 7740 was released as FUNIP-2. In Peru, composite J is grown on about 1,000 ha as Opaco Huascaran, and the national program is about to release another QPM variety from Population 63.

In Africa, Senegal released the varieties Obregón 7740 and Poza Rica 8362 and Ghana released years ago La Posta-O_2 variety (soft endosperm) and Crop Research Institute-Global 2000 evaluated in 1990 a large number of hard endosperm QPM germplasms.

In Asia, China has the most active QPM breeding program. An open-pollinated variety derived also from Tuxpeño-O_2, was released in Guangsi Province in the south of China, and soft-endosperm, opaque-2, single-cross hybrids are being grown in the country's northern temperate regions. Vietnam has released a

variety derived from Population 63. Scientists in
India and the Philippines also have been evaluating
CIMMYT's QPM germplasm.

<u>Conclusions</u>. CIMMYT now has available QPM gene pools,
populations and open-pollinated experimental varieties
and experimental hybrids of early and late maturity,
with white and yellow grain, and tropical, subtropical
and tropical highland adaptation. The grain type of
this germplasm is often indistinguishable from that of
normal maize, except that it has superior protein
quality. This QPM germplasm provides an excellent
base for further work either in population
improvement, or hybrid development for any breeding
program interested in the production of maize with
superior nutritional value.

LITERATURE CITED

Arriaga, C., E.L. Deckard, R.L. Lambert, and R.H.
    Hageman. 1970. Grain protein and lysine content
    of two normal corn hybrids and their opaque-2
    derivations as affected by supplemental nitrogen
    applications. Agronomy Abstr., p. 74, ASA,
    Madison, WI.
Bjarnason, M. 1990. El programa de maíz con calidad
    de proteina del CIMMYT: estado presente y
    estrategias futuras. CIMMYT Apdo. Postal 6-641,
    Mexico, D.F. Mexico.
Bjarnason, M., W.G. Pollmer, and D. Klein. 1976.
    Inheritance of modified endosperm structure and
    lysine content in opaque-2 maize. I. Modified
    endosperm structure. Cereal Res. Commun. 4:401-
    410.
Bjarnason, M., K. Short, S.K. Vasal, and E. Villegas.
    1988. Genetic improvement of various quality
    protein maize (QPM) populations. Agronomy Abstr.,
    p. 74, ASA, Madison, WI.
CIMMYT. 1970. Annual Report 1968-69. Mexico D.F.
CIMMYT. 1990. CIMMYT International Testing Program,
    1988 Final Report. CIMMYT, Mexico, D.F.

Dudley, J.W. 1974. (ed.). Seventy generations of selection for oil and protein in maize. Crops Sci. Soc. of Amer., Madison, WI.

Francis, C.A., L.E. Alvarez, V.D. Sarria, and P. Pinstrup-Andersen. 1972. Yields and acceptability of opaque-2 maize in the tropics of Colombia. Agronomy Abstr., p. 190, ASA, Madison, WI.

Frey, K.J. 1951. The interrelationships of proteins and amino acids in corn. Cereal Chemistry 28:123-132.

Gevers, H.O. 1972. Breeding for improved protein quality in maize. Trans. R. Soc. S. Afr. 40(2):81-92.

Glover, D.V., P.L. Crane, P.S. Misra, and E.T. Mertz. 1975. Genetics of endosperm mutants in maize as related to protein quality and quantity. p. 228-240. In: High Quality Protein Maize. Hutchinson, Ross Publishing Co., Stroudsburg, PA.

Glover, D.V. and E.T. Mertz. 1987. Corn. p. 183-336. In: Nutritional Quality of Cereal Grains: Genetic and Agronomic Improvement. R.A. Olson and K.J. Frey (eds.) ASA Monogr. 28. ASA, Madison, WI.

Harpstead, D.D. 1969. High lysinc maize in its proper perspective. p. 74-80. In: 24th Ann. Corn and Sorghum Res. Conf., American Seed Trade Assoc., Chicago, IL.

Hernandez, H.H. and S.L. Bates. 1969. A modified method for rapid tryptophan analysis in maize. CIMMYT Res. Bul. No. 13. May 1969.

Ingversen, J., B. Koie, and H. Doll. 1973. Induced seed protein mutant of barley. Experimentia 29:1151-1152.

Jiménez, J.R. 1966. Protein fractionation studies of high lysine corn. In: Proceedings of the High Lysine Corn Conference. p. 74. Corn Ind. Found., Washington, D.C.

Lambert, R.J., D.E. Alexander, and J.W. Dudley. 1969. Relative performance of normal and modified protein (opaque-2) maize hybrids. Crop Sci. 9:242-243.

McWhirter, K.S. 1971. A floury endosperm high-lysine locus on chromosome 10. Maize Genet. Coop. Newsletter 45.

Mertz, E.T., L.S. Bates, and O.E. Nelson. 1964.
Mutant gene that changes protein composition and
increases lysine content of maize endosperm.
Science 145:279-280.

Mertz, E.T. 1974. A rapid ninhydrin color test for
screening high lysine mutants of maize, sorghum,
barley and other cereal grains. Cereal Chem.
51:304-307.

Miller, R.C., L.W. Aurand, and W.R. Flach. 1950.
Amino acids in high and low protein corn. Science
112:57-58.

Mossberg, R. 1969. Evaluation of protein quality and
quantity by dye-binding. In: New approaches to
breeding for improved plant protein. IAEA/FAO,
STI/Pub 212. Vienna. p. 151-161.

Munck, L., K.E. Karlsson, and A. Hagberg. 1969.
Selection and characterization of a high-protein,
high-lysine variety from the world barley
collection, presented at the Second International
Barley Genetics Symposium, Pullman, WA., USA, July
6-11.

Nelson, O.E., E.T. Mertz, and L.S. Bates. 1965.
Second mutant gene affecting the amino acid pattern
of maize endosperm proteins. Science 150:1469-
1470.

Nelson, O.E. 1966. Opaque-2, floury-2 and high
protein maize. p. 156-160. In: Proc. High Lysine
Corn Conference. E.T. Mertz and O.E. Nelson (eds.)
West Lafayette, IN. Corn Industries Research
Foundation, Washington.

Ortega, E.I. and L.S. Bates. 1983. Biochemical and
agronomic studies of two modified hard-endosperm
opaque-2 maize (*Zea mays L.*) populations. Cereal
Chem. 60:107-111.

Paez, A.V. 1973. Protein quality and kernel
properties of modified opaque-2 endosperm corn
involving a recessive allele at the sugary-2 locus.
Crop Sci. 13:633-636.

Paez, A.V., M.S. Zuber. 1973. Inheritance of test-
weight components in normal, opaque-2, and floury-2
corn (*Zea mays L.*). Crop Sci. 13:417-419.

Paez, A.V., J.L. Helm, and M.S. Zuber. 1969. Lysine content of opaque-2 maize kernels having different phenotypes. Crop Sci. 9:251-252.

Paez, A.V., J.P. Ussary, J.L. Helm, and M.S. Zuber. 1969. Survey of maize strains for lysine content. Agron. J. 61:886-889.

Paulis, J.W., J.S. Wall, and W.F. Kwolec. 1974. A rapid turbidimetric analysis for zein in corn and its correlation with lysine content. J. Agr. Food Chem. 22:313-317.

Phillips, R.L., P.R. Morris, F. Wold, and B.G. Gegenbach. 1981. Seedling screening for lysine-plus-threonine resistant maize. Crop Sci. 21:601-607.

Pinstrup-Andersen, P. 1971. The feasibility of introducing opaque-2 maize for human consumption in Colombia. Tech. Bull. 1. CIAT, Cali, Colombia, South America.

Poey, F.R. and E. Villegas. 1972. Modified endosperm phenotype of opaque-2 maize in relation to genotype and protein value. Agr. Abstr. ASA p. 70. Madison, WI.

Singh, J. and V.L. Asnani. 1975. Present status and future prospects of breeding for better protein quality in maize through opaque-2. In: High-Quality Protein Maize. p. 86-99. Hutchinson Ross Publishing Co., Stroudsburg, PA.

Singh, R. and J.D. Axtell. 1973. High lysine mutant gene (hl) that improves protein quality and biological value of grain sorghum. Crop Sci. 13:535-539.

Sreeramulu, C. and L.F. Bauman. 1970. Yield components and protein quality of opaque-2 and normal diallels of maize. Crop Sci. 10:262-265.

Tanksley, S.D.; N.D. Young, A.H. Patterson, and M.W. Bonierbale. 1989. RLFP mapping in plant breeding: new tools for an old science. Biotechnology 7:257-264.

Tsai, C.V., L.W. Hansel, and O.E. Nelson. 1972. A colorimetric method of screening maize seeds for lysine content. Cereal Chem. 49:572-579.

Vasal, S.K. 1975. Use of genetic modifiers to obtain normal-type kernels with the opaque-2 gene. In:

High Quality Protein Maize. p. 197-215.
Hutchinson Ross Publishing Co., Stroudsburg, PA.
Vasal, S.K. 1986. Approaches and methodology in the
development of QPM hybrids. Anais do XV Congresso
National de Milho e Sorgo, Brasilia. EMBRAPA-
CNPMS, Documentos 5. p. 419-430.
Vasal, S.K., E. Villegas, M. Bjarnason, B. Gelaw, and
P. Goertz. 1980. Genetic modifiers and breeding
strategies in developing hard-endosperm opaque-2
materials. In: Improvement of quality protein
traits of maize for grain and silage use. p. 37-
71. W.G. Pollmer and R.H. Philipps (eds.) Wijhoff,
The Hague.
Vasal, S.K., E. Villegas, and C.Y. Tang. 1984a.
Recent advances in the development of quality
protein maize germplasm at the Centro Internacional
de Mejoramiento de Maíz y Trigo. p. 167-189. In:
Cereal Grain Protein Improvement, IAEA, Vienna.
Vasal, S.K., E. Villegas, C.Y. Tang, J. Werder, and
M.Read. 1984b. Combined use of two genetic
systems in the development and improvement of
quality protein maize. Kulturpflanze 32:171-185.
Villegas, E., E.I. Ortega, and R. Bauer. 1984.
Chemical methods used at CIMMYT for determining
protein quality in cereal grains. Centro
Internacional de Mejoramiento de Maíz y Trigo,
Mexico, D.F. Apdo Postal 6-641, Mexico.
Zuber, M.S. and J.L. Helm. 1975. Approaches to
improving protein quality in maize without the use
of specific mutants. p. 214-252. In: High Quality
Protein Maize. Hutchinson Ross Publishing Co.,
Straudsburg, PA.

DEVELOPMENT OF MODIFIED OPAQUE-2
MAIZE IN SOUTH AFRICA

H.O. Gevers and J.K. Lake
Grain Crops Research Institute,
c/o University of Natal,
P.O. Box 375, Pietermaritzburg, 3200.

INTRODUCTION

When the beneficial nutritional effects of the opaque-2 gene were first reported by Mertz, Bates & Nelson (1964), over 60% of all maize in South Africa was consumed by humans, largely as a staple diet of white maize. Then, and for many years thereafter, maize provided not only the main energy source but also a large proportion of the dietary protein in maize eating communities. This situation was conducive to the occurrence of deficiency diseases kwashiorkor and pellagra which, respectively, are related to the nutritionally poor source of lysine and tryptophan in zein, the main endosperm protein of maize. Obviously, a genetic improvement in these and other protein constituents of maize, which the recessive opaque-2 mutant offered, would not only counteract these deficiencies but would also contribute greatly to a generally improved nutritional level of the human consumer of maize and maize products. These were important considerations when the high lysine maize (HLM) breeding work commenced in 1965.

The use of nutritionally improved maize in animal feed, mainly for monogastrics, likewise, had considerable economic potential if it could reduce the proportion of widely used and costly dietary components such as fish meal. These economic benefits would logically be at a maximum if HLM were to attain yields and other agronomic attributes that were competitive with those of conventional or normal maize. At first, this possibility seemed remote as the undesirable agronomic effects associated with this mutant, namely, a lower yield, softer kernel and greater susceptibility to kernel and storage diseases, indeed seemed forbidding in overcoming these weaknesses by breeding and selection. However, this was the only possible course of action that offered any hope of reaching this seemingly unattainable nutritional and economic potential in the future.

The fact that in 1965 this country had only recently introduced hybrid maize made the conversion of inbreds and hybrids rather than populations to this mutant the logical starting point in an HLM breeding program. This would automatically lead to the development of more productive hybrids for the progressive farmer and, at the same time, open pollinated varieties or synthetics for the poorer farming communities.

This paper outlines the progress made in the development of modified opaque-2 maize and the introduction of commercial HLM in South Africa.

MATERIALS AND METHODS

Reports that appeared soon after the first announcements on the opaque-2 mutant (Mertz et al 1964), confirmed that considerable genetic variation in the expression of this gene was present in segregating populations (Alexander, 1966; Alvey & Hamilton,1967). The variation in the expression of this gene was the underlying basis of all the selection strategies employed in the development of HLM. Further, successful development and future use of HLM was dependent on the mass backcross conversion to opaque-2 of established and highly selected normal breeding material. The task was formidable because,

in the beginning, no opaque-2 breeding material
existed whilst more complicated and costly screening
methods would have to be employed in maize breeding
programs that were otherwise similar to conventional
ones. Although the main effect is that of a single
gene, the selection procedures designed to modify its
expression were undoubtedly based on quantitative
gene action (Gevers, 1972; 1989). Selection
strategies were based on the following main
criteria:-

<u>Relative kernel mass (RKM)</u>. Initially, the RKM of
opaque-2 segregants, as compared with their normal
counterparts, was the most important selection
criterion in counteracting the lower yield associated
with this mutant. Opaque-2 segregants with the
highest RKM were chosen both for the next backcross
and as new lines.

<u>Visual screening and selection</u>. Opaque-2 kernels
selected on the basis of RKM were all visually
screened on illuminated ground glass screens and
routinely compared with their normal counterparts.
Opaque-2 kernel segregants exhibited great
differences in translucence, varying from that of the
undesirable and soft true opaque-2 which, as the name
implies, allows no passage of light through the
kernel, to that of the desirable translucence of the
phenotypically normal counterpart. Selection for
both RKM and translucence was expected to accelerate
the accumulation of genetic modifiers in breeding
material being converted to opaque-2. The routine
visual pre-selection of modified opaque-2 ears in
breeding nurseries at harvest was an important
additional procedure. The most desirable modifier
complexes would be those that brought about a close
resemblance of the normal kernel in terms of hardness
and translucence and preferably a minimal decline in
lysine content.

<u>Biochemical analysis</u>. Endosperm material was
continuously monitored for protein and lysine content
by standard Kjeldahl procedures and Beckman amino

acid analyzer, respectively. In later years, biochemical analysis was confined to whole grain material. Lysine content was expressed both as percentage of protein and of sample. For routine sample analysis only the amino acid lysine was determined according to an abbreviated Beckman short column procedure developed locally by Dennison (1971). Complete amino acid analyses were carried out only in specialized nutritional and feeding studies mindful that, in addition to lysine, a higher tryptophan and lower leucine content of the maize endosperm are other beneficial nutritional changes brought about by the opaque-2 gene.

<u>Kernel density and hardness.</u> It was also of interest to determine kernel density, which would presumably be positively associated with kernel hardness and be lower for opaque-2 segregants. This was periodically determined by standard methods for comparable normal and opaque-2 breeding material and cultivars. Using a Technicon InfraAlyzer 450, kernel hardness was determined by means of a Near Infrared Reflectance (NIR) spectrophotometer, which was calibrated according to kernel hardness data obtained by the method of Vorwerck & Miecke (1973).

<u>Electrophoretic patterns</u>. Lastly, selected samples of normal, modified opaque-2 and soft opaque-2 genotypes were subjected to an analysis of electrophoretic zein patterns in kernel endosperms according to a procedure recently used by Wallace et al (1989). According to this method, these kernel variants exhibit characteristic quantitative differences in content and electrophoretic distribution of gamma and alpha zein and may therefore be used as criteria in assessing progress by selection for modified (hard) opaque-2 kernel endosperms.

<u>Genetic evaluation</u>. More critical evaluation of breeding progress was made possible by the study of the relative performance of modified opaque-2 inbreds and their near-isogenic, normal counterparts in test-

cross combinations with a common set of opaque-2 testers and isogenic tester pairs. Further, estimates of general and specific combining ability for kernel hardness, according to Griffing (1956), provided predictions of possible future selection progress for this important characteristic.

<u>Yield evaluation</u>. The regular yield and agronomic evaluation of highly selected experimental opaque-2 hybrids and breeding material was the final and critical step in assessing the selection progress made and the economic potential of this new type of maize.

<u>Nutritional evaluation</u>. Periodic feeding trials were carried out co-operatively with locally developed opaque-2 maize by other research workers with monogastric animals and ruminants for the purpose of monitoring nutritional responses and to confirm results obtained elsewhere.

RESULTS AND DISCUSSION

<u>Selection progress</u>. The problem of a low RKM in segregating opaque-2 material is illustrated in Fig.1 where the distribution of this ratio is shown for a population undergoing conversion to opaque-2. Obviously, the breeder would concentrate on the 10 % of the population showing a yield deficit of less than four percent. Similarly, inbred lines, such as I137TN, with relatively large numbers of segregants with a high RKM (Fig. 2) were selected for future use and poorer ones discarded, thus ensuring the early elimination of breeding material unsuitable for conversion to opaque-2 (Gevers, 1972). The aim of selection, throughout, was to attain the same or better yield and agronomic performance as the normal counterpart.

The potential for successful selection against a lower yield in opaque-2 breeding material was thus indicated and these differences ascribed to genetic modifiers which change the expression of the mutant

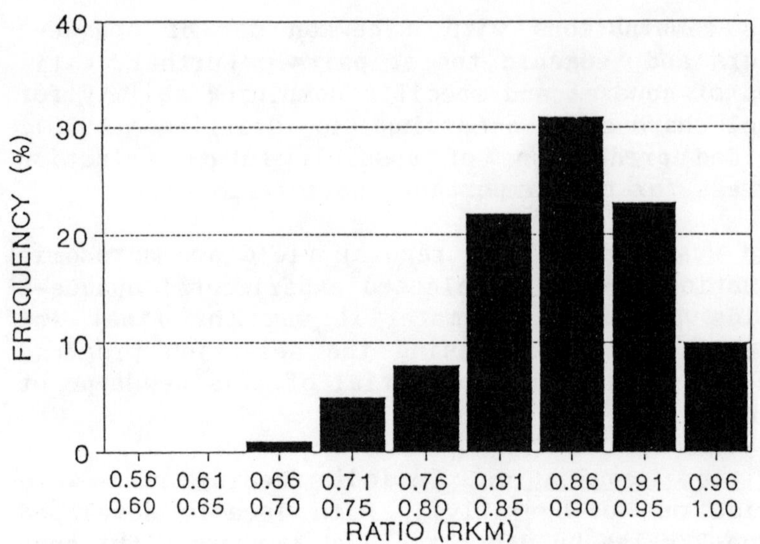

Fig.1. Distribution of RKM in a maize population segregating for opaque-2.

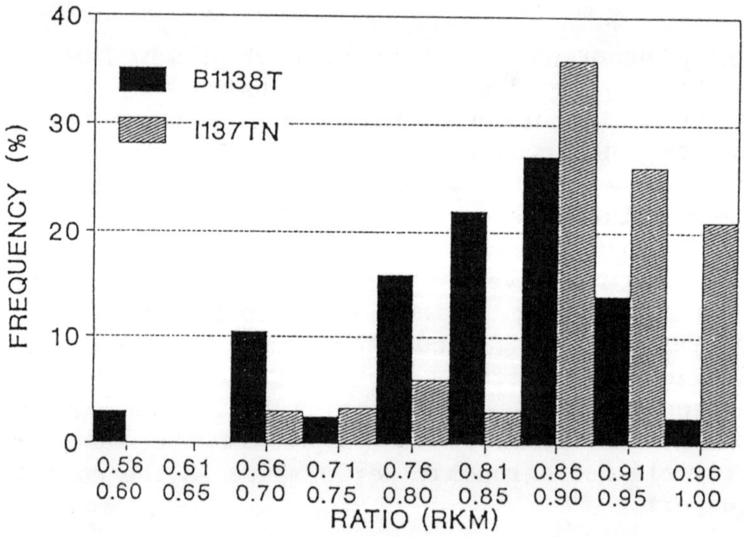

Fig.2. Distribution of RKM in segregating backcross populations of two inbred lines being converted to opaque-2.

gene. Although more precise inheritance studies were to be carried out in the future, these preliminary findings generally agree with those of other workers and were the most important basis of the selection strategy in opaque-2 conversion programs.

Specifically, these findings are consistent with those of workers at CIMMYT in Mexico, who found that the inheritance of kernel vitreousness (assumed to be synonymous with hardness) in opaque-2 maize was largely based on additive gene action (Vasal et al 1980). Recurrent selection for this character should therefore approach the yield level and kernel attributes of normal maize within a relatively short time. It remained to concentrate relentlessly on the selection for modifiers in variable genetic backgrounds by means of RKM, kernel hardness and density, which seemed to be highly correlated.

<u>Local M and F backgrounds</u>. During the intensive program of conversion and selection, a number of genetic backgrounds emerged as agronomically superior, while other elite and proven normal kernelled ones were unacceptable. Of particular significance were the genetically divergent local M and F heterotic backgrounds, which have played a major role in the breeding progress made (Gevers, 1979). The situation, nevertheless, was that the germplasm resources available in conventional maize breeding were much larger than those in opaque-2 programs which, initially, would possibly retard breeding progress. This new material was the basis of the future progress made in HLM breeding.

<u>First hybrids</u>. <u>White high lysine maize</u>. The first major evaluation of experimental opaque-2 maize hybrids was in 1976-77, 11 years after the breeding program was initiated. HL1, the first commercial white high lysine hybrid, was released in 1979. A year after its release, HL1 was ranked third out of 49 commercial maize hybrids for yield in the 1980/81 National Maize Cultivar Trials carried out over more than 30 localities in South Africa (Gevers, 1982).

It was thus highly competitive with the highest yielding commercial hybrids and superior to the majority of the widely used and popular hybrids of that period. However, it was not readily accepted by the millers and manufacturers of white maize meal for the reason that its kernel was too soft and milling wastage too high.

<u>Dry milling requirements</u>. In a study of the physical properties of HLM, it was noted that maize used for human consumption is largely associated with the dry but also with the wet milling industry (Van Twisk et al 1976). Any departure from the physical kernel properties of maize cultivars normally used in these specialized industries, would possibly require a reassessment or modification of milling procedures. Any change in kernel properties would therefore more likely be viewed with disfavor despite the improved nutritional value it offered.

Although numerous minor differences in physical and chemical properties were found in this relatively unselected opaque-2 material, no critical disadvantages of this new maize type were detected in the wet milling processes. However, the softness of the kernel and associated kernel properties of a low density, loosely packed starch grains in the endosperm and relatively low proportion of hard (horny) endosperm, were distinct and serious disadvantages in the dry milling industry (Van Twisk et al 1976). These poor physical kernel properties would apparently interfere with the highly desirable granulation characteristics of conventional hard-kernelled white maize, which is so successfully utilized in the manufacture of the special, sifted granulated maize meal, maize rice and samp, which are particularly popular among the large maize eating communities. The considerable loss of nutritional value in these maize products in the dry milling process is a result of refinement, and its possible serious nutritional consequences to the consumer have been highlighted (Quicke, 1982).

Industrially, no progress in making available such a nutritionally superior product and staple seemed possible without further genetic improvement of white HLM maize in terms of kernel hardness unless some innovative adjustments in the industrial milling procedures were introduced to accommodate a softer kernel. Already there is a noticeable decline in the consumption of maize meal in South Africa in favour of bread and certain cereal foods that lend themselves better to easy and rapid meal preparation. If indeed this is a factor, which is likely among the large maize eating population, then one can visualise the industrial potential of an easily prepared, instant and nutritious food based on white HLM.

Kernel hardness. Recent results, presented in Tables 1 & 2, outline the present status of experimental white modified opaque-2 hybrids in terms of yield and kernel hardness and genetic variances derived in diallel analyses for elite breeding material of this type, respectively. The latter are preliminary data.

First indications are that a number of new white high lysine hybrids have a kernel hardness considerably higher than the first-released white opaque-2 hybrid HL1 and also the commercial white hybrid SR52, which has the lowest acceptable kernel hardness for industrial processing (Table 1). These experimental white high lysine hybrids also have higher figures than numerous commercial white hybrids in wide usage in South Africa. Moreover, preliminary estimates of general combining ability (Gca) mostly indicate that further selection progress in this material is possible and that, potentially, the kernel hardness of conventional maize may be fully attained in the future (Table 2). Relatively high estimates of specific combining ability (Sca), which are expected in such highly selected material, nevertheless indicate that non-additive effects are also of importance in the inheritance of this character. Undoutedly, therefore, these new white modified opaque-2 hybrids have the potential for the successful industrial processing of this maize type.

TABLE 1

Relative Yield and Kernel Hardness (%) for White
Mod. O_2 Maize Hybrids and Controls(C)(Summary)

Rank	Hybrids	Relative Yield	Hardness %
1	(BO46W.BO59W)FO280W	115.1	62.7
2	(FO215W.FO197W)RO452W	114.9	65.9
3	(KO54W.FO197W)RO452W	113.6	64.4
4	(FO215W.FO197W)RO460W	113.3	68.1
5	(FO215W.KO54W)RO452W	113.0	67.1
6	(FO215W.FO197W)RO469W	112.1	64.3
7	(FO215W.FO197W)RO445W	111.9	65.0
8	(FO215W.FO197W)RO408W	111.8	63.7
9	(FO215W.FO197W)RO463W	111.3	62.6
10	(FO215W.FO197W)RO464W	111.3	66.5
11	(FO215W.FO197W)RO406W	109.6	65.4
12	PNR 6549 C N	109.6	66.1
13	RO465W X BO89W	109.3	68.1
14	POWS X RO452W	108.9	67.1
15	A1849W C N	108.7	67.6
16	(FO215W.FO197W)RO409W	108.3	67.3
17	HL 8 C	108.1	58.1
18	(FO215W.FO197W)RO466W	107.8	67.4
19	RO500W X BO89W	107.4	63.9
20	(FO215W.FO197W)BO59W	105.9	63.6
45	HL 1 C	96.4	63.7
51	SNK 2147 C N	90.3	63.9
60	SR 52 C N	83.7	62.8

N= Normal Kernel

These findings are in full agreement with those
of Vasal et al (1980) at CIMMYT. It is long overdue
that the maize milling industry makes available these
nutritional benefits to the consumer of maize in all
walks of life, but especially the poorer communities.
Obviously, these new hybrids and populations will in
time also be available for home grown consumption in
these communities.

58

TABLE 2

Kernel Hardness(%), Gca and Sca Variances and Lysine Content of Parents
for an 8-line White Maize Diallel (Preliminary Data)

Inbred	RO520W	RO454W	RO588W	FO197W	FO284W	FO215W	KO9W	BO89W	Variance		Lysine
									Gca	Sca	% Prot.
RO520W		63.1	64.2	62.9	63.9	61.7	62.0	60.0	0.86	1.44	4.09
RO454W			65.5	65.8	65.3	65.4	60.9	63.3*	0.99	1.45	3.74
RO588W				63.9	62.7	65.0	62.2	62.6	0.20	0.69	3.73
FO197W					62.2	65.8	61.8	64.0	0.27	1.18	3.98
FO284W						63.3*	63.2	62.6	0.00	1.69	3.67
FO215W							66.4	62.3	1.19	0.96	4.06
KO9W								61.3	0.87	3.01	3.55
BO89W									1.44	0.70	4.17
Means	62.5	64.2	63.7	63.8	63.3	64.3	62.5	62.3			

* Missing plot values

Yellow high lysine maize. Yellow HLM, which would be used mainly in animal feeds, where kernel hardness is not a serious problem, became the most urgent breeding objective in later years. In 1982, HL2, the first South African yellow high lysine hybrid, was released commercially. At the time much more assured of success than white hybrids, HL2 was agronomically highly acceptable and, at the time, competitive with the best commercial hybrids (Gevers, 1986). Its modified opaque-2 grain type is hardly distinguishable from other commercial hybrids and definitely superior to those normal yellow hybrids with apparent large dosages of Corn Belt material.

Intense selection for modifiers has been highly successful in making HLM agronomically competitive and acceptable. HL2 is agronomically superior to the white commercial hybrid HL1. At no stage of selection could a decline in lysine content be detected.

Yield performance data from at least 11 localities for the most advanced HLM hybrids and normal-kernelled controls for the past four seasons are summarized in Fig. 3. The commercial yellow control hybrid PNR 6528 was chosen on the basis of its outstanding yield record over five seasons in Co-operative Maize Trials of the Natal maize breeding program, National Phase 2 Cultivar Trials and trials by the Pioneer Seed Company who released it (Gevers, 1989).

Accurate yield trial data for four consecutive seasons clearly show that, in comparison with this control, the performance of the highest yielding high lysine hybrids represent noteworthy breeding progress. These hybrids are competitive with high yielding and widely used commercial maize hybrids in South Africa. In particular, the newly released commercial high lysine hybrid HL8 has maintained the same yield as PNR 6528 over these seasons, while new experimental ones have made advances on this hybrid (Fig. 3).

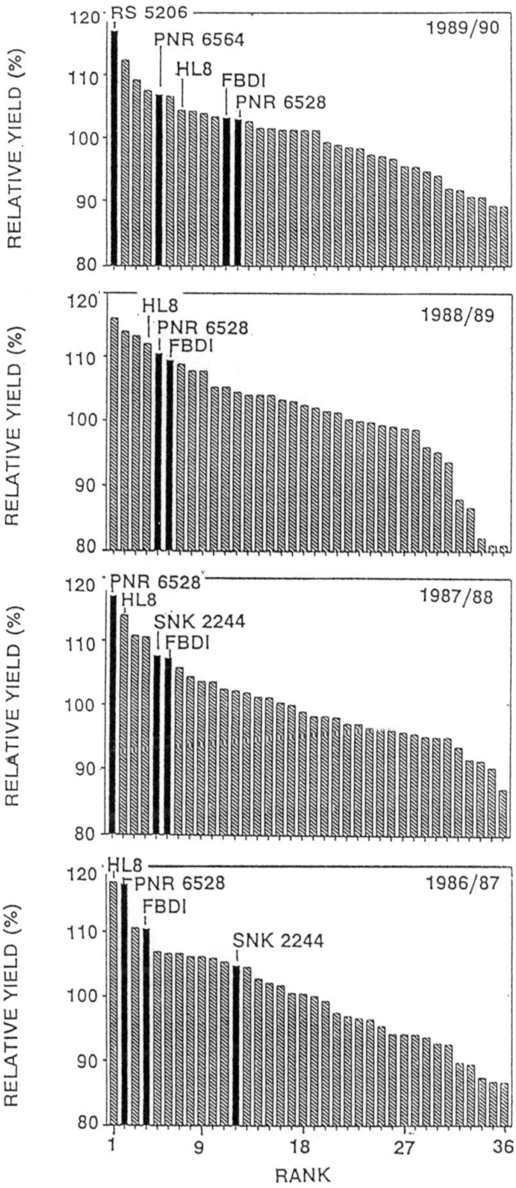

Fig.3. Mean relative yields of the best high lysine
(barred) and normal (solid) maize hybrids over four
seasons.

In the most recent season (1989/90), a high yielding commercial hybrid RS5206 (or ZS206) has nevertheless had the highest mean yield ranking. However, it was found that it was significantly higher yielding than high lysine hybrids in only two out of 13 localities, indicating continued competitiveness of the latest high lysine hybrids (unpublished Progress Report, 1989/90).

<u>Other advances</u>. The advances made in breeding HLM for yield also apply to other important agronomic characteristics and criteria used in monitoring the selection progress. These are listed below.

<u>Disease resistance</u>. Numerous results confirm that most of the advanced high lysine hybrids have markedly lower ear rot ratings than the majority of commercial hybrids in use today (Gevers, 1986). A specific example is that the hybrid HL2, which also conforms to high agronomic standards, had the lowest ear rot rating of 49 commercial maize hybrids included in the National Phase 2 Cultivar Trials of 1986/87 (Maize Board, 1988). This applies in large measure to new high lysine maize cultivar releases.

<u>Kernel density and hardness</u>. Most promising new high lysine hybrids now have a kernel density equal or better than representative normal cultivars, again indicating that the selection for modifier genes has been highly effective (Fig.4). Further, kernel density may be positively associated with kernel hardness as other data suggest (Gevers, 1989). Indeed, considerable improvements have been made on HL1, which in previous estimates had a hardness of only 44% and which was unacceptable to industrial processors of white maize. As shown previously in this paper (Tables 1 & 2), considerable progress has recently been made in this regard. Modified white opaque-2 maize now has an equal or greater kernel hardness than the minimum level required by industrial processors of white maize meal.

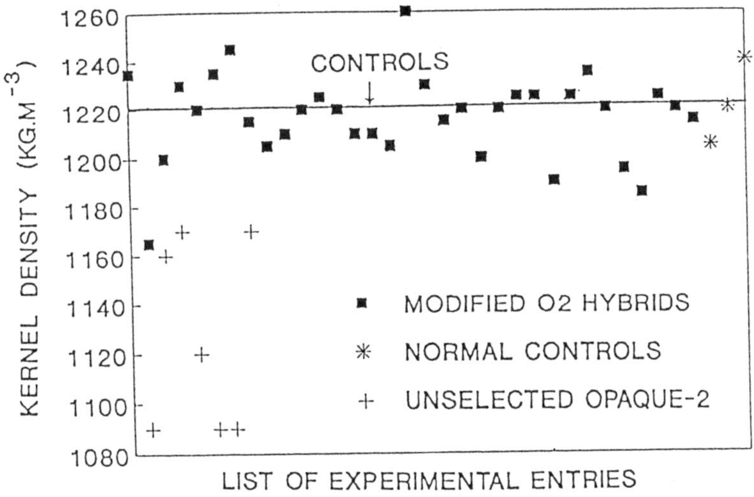

Fig.4. Kernel density (kg m^{-3}) for elite modified opaque-2 maize hybrids normal controls (mean indicated) and unselected opaque-2 segregants.

<u>Performance of near-isogenic parents</u>. Of further interest is that elite modified opaque-2 inbred lines and their near-isogenic, normal counterparts, with few exceptions, have given similar yields in crosses with opaque-2 and normal testers (Gevers & Lake, 1990a) (Tables 3 & 4). However, as has been the general experience, certain genotypes are less suitable for conversion to opaque-2, as exemplified by the I137TN group, where the opaque-2 recoveries are usually significantly lower yielding in testcrosses. Softer dent kernel types such as these, which include Corn Belt breeding material, have generally been poorer converters to opaque-2.

In all other cases the differences in test cross yields of the modified opaque-2 inbreds and their near-isogenic counterparts are usually not significantly different and, in some cases, the performance of modified opaque-2 testcrosses is consistently superior (Tables 3 & 4). In these comparisons there is therefore no longer a consistent difference among many elite and highly selected near-isogenic opaque-2 and normal maize genotypes. The

TABLE 3

Performance of Near-Isogenic Normal (N) and Opaque-2
(O) Inbred Groups with Modified O_2 Testers

Isogenic Group		No. of Tests	Relative Yield*		
			N > O	O > N	O = N
A. 1137TN	N				
BO404Y	O	6	5		1
BO412Y	O	6	6		
KO439Y	O	6	4		2
B. D940Y-1	N				
DO940Y-1	O	7	2		5
C. K130Y	N				
KO429Y	O	6	2	1	3
D. B312Y	N				
BO310Y	O	5		2	3

* Comparisons based on means of 2 locations

TABLE 4

Yields (t/ha) of Near-Isogenic Normal(N) and Modified
Opaque-2(O) Inbred Groups with N and O Testers.

Isogenic Group		Isogenic Testers		
		B312Y (N)	BO310Y (O)	Means
A 1137TN	N	6.6	6.7	6.6
BO404Y	O	6.0**	6.0**	6.0
BO412Y	O	6.2*	5.9**	6.0
KO439Y	O	6.3	6.1**	6.2
B D940Y-1	N	6.5	6.7	6.6
DO940Y-1	O	6.3	6.4	6.4
C K130Y	N		6.2	6.2
KO429Y	O	6.0	6.1	6.0

Approx. statistical differences (*P=5%, **P=1%)
indicates lower yielding opaque-2 variant in group.

implications of these findings are that, potentially, highly selected modified opaque-2 genotypes need not have any yield disadvantage, and should thus allow the full commercial exploitation of their superior nutritional value without economic risk.

<u>Electrophoretic zein patterns</u>. The analysis of SDS electrophoretic patterns for the endosperm proteins of normal, modified (hard) opaque-2 and (soft) opaque-2 genotypes by an elegant new technique reported by Wallace et al (1989), has provided further evidence of progress in selecting for modified opaque-2 kernel types in maize.

A selection of local populations and inbreds, which had previously only been visually classified, were shown to conform closely to the postulate that modified opaque-2 genotypes (HEO2) with high levels of lysine and tryptophan, traditional appearance and high yields, characteristically had a markedly higher gamma-zein (Mr 27kD) and minimal alpha-zein (Mr-22kD) component than both normal(N) and ordinary soft (SO,) maize genotypes (Table 5). Most of these HEO2 inbred lines are represented in the highest yielding and agronomically superior high lysine hybrids released in South Africa so far. It is of further interest that the CIMMYT HEO2 populations WHEO2 and Amarillo Bajio have zein patterns that closely resemble that of local HEO2 population POWS, which has been shown to be a good combiner and source of inbreds (Table 5). The local work thus finds a good parallel in the results obtained at CIMMYT.

<u>Amino acid levels</u>. The levels of protein, lysine and tryptophan in maize endosperms suggest that there is little or no decline of these constituents in modified opaque-2 variants.(Table 5). Data for near-isogenic inbred and population pairs, taken from Table 5 and elsewhere, further highlight the striking differences in the nutritionally important amino acids lysine and tryptophan in isogenic N and HEO₂ genotypes which, however, have a comparable kernel hardness (Table 6). And, as shown previously for the

Electrophoretic Zein Patterns of Normal (N) and
Opaque-2 (O) Maize Breeding Material

Zein Pattern	Pedi-gree	Type	Kernel Type	Prot %	Lys** %	Trypt %
	WHEO2	Pop.	HEO2*	9.3	3.5	0.8
	NPPES1		N	10.1	2.1	0.4
	POWS		HEO2	9.2	3.5	0.8
	Amar.Bajio		HEO2	10.7	3.0	0.7
	KO326Y	Inbred	HEO2	8.3	4.0	0.9
	DO940Y.523		HEO2	7.9	4.0	0.9
	DO383Y		SO2	9.6	2.6	0.6
	KO437Y		HEO2	7.9	4.4	1.0
	KO439Y		HEO2	5.7	3.2	0.7
	DO620Y		HEO2	9.2	3.5	0.8
	BO385Y		HEO2	8.3	3.2	0.7
	KO315Y		HEO2	8.8	3.2	0.7
	D940Y-1		N	9.8	1.5	0.3
	I137TN		N	10.7	1.5	0.3

* HEO2 = Hard endosperm opaque-2, N = Normal,
SO2 = Soft opaque-2. ** Lysine and tryptophan
expressed as percent of protein

TABLE 6
Electrophoretic Zein Patterns of Near Isogenic Maize
Inbred Pair D940Y-1(N) and DO620Y(O) and
Populations NPPES1 (N) and POWS (O)

Zein Pattern	Pedi-gree	Kernel Type	Prot %	Lys** %	Trypt %	Hardness %
	D940Y-1	N	9.8	1.5	0.3	59.0
	DO620Y	HEO2*	9.2	3.5	0.8	58.4
	NPPES1	N	10.1	2.1	0.4	64.9
	POWS	HEO2	9.2	3.5	0.8	61.6

*/** See footnote to Table 5.

comparable isogenic inbred pair D940Y-1 and DO940Y-1 (Tables 3 & 4), they are usually at competitive yield levels. The study of electrophoretic zein patterns has therefore not only assisted in highlighting the nutritional and physical differences in N and HEO_2 genotypes, but is a valuable tool in monitoring the progress made in HLM breeding.

<u>Competitive yields</u>. The yield data reported in the previous sections can hardly be regarded as representative and applicable to all maize growing areas of this country. However, indications are that promising new high lysine hybrids are at highly competitive yield levels, as compared with the best normal commercial hybrids. They may be grown without any agronomic or economic risk in large parts of South Africa, particularly in the higher rainfall areas. The positive results obtained by sustained selection and breeding research has indicated a potential, namely, that modified opaque-2 hybrids with yields and agronomic qualities as good or better than conventional maize are attainable. Further, recent diallel analysis of improved populations indicate that elite new modified opaque-2 populations gave the highest yields in crosses and therefore offer the scope for continued further breeding progress (Gevers, 1990b).

<u>Area planted to high lysine hybrids</u>. After a slow beginning, the area planted to HL2 and the latest hybrid HL8, after many preliminary production problems, has steadily increased from 235 ha in 1983 to about 20 000 ha in 1988, and an anticipated 40 000 ha in 1990/91. Were it not for a seed shortage and other initial problems, the area would have been at least doubled in 1990. The choice of high lysine hybrids by farmers is in many cases based solely on its superior agronomic characteristics and relatively high resistance to the <u>Diplodia</u> ear rots. The successful industrial exploitation of the superior nutritional value of this new maize type is therefore entirely feasible. Despite serious initial and remaining resistance to HLM, at least four commercial

seed companies are in the process of placing high
lysine hybrids on their sales lists.

<u>Nutritional and economic implications</u>. <u>The human
consumer</u>. The nutritional problems associated with
a staple diet of conventional (normal) maize relate
to the relatively poor lysine and tryptophan
deficient maize protein zein. The higher biological
value and potential nutritional benefits of HLM
therefore basically lie in a more balanced protein as
represented by a high lysine and tryptophan and,
indirectly, to a reduced leucine content.

Although these nutritional improvements in
opaque-2 maize, including that of an improved source
of niacin, have been demonstrated in a variety of
laboratory and domestic animals and humans, its
beneficial contribution to human nutrition and well
being is best illustrated in the clinical treatment
of malnourished young-children (Quicke, 1982;
National Research Council, 1988; Bressani, 1975)
(Table 7 & Fig. 5). It is a measure of the
nutritional value of HLM that it is equated to milk
or casein in these reports, but its value and
potential are much more far-reaching. Further, these
beneficial effects are not confined to children, but

TABLE 7

Mean N-balance Index (NBI) Values for Normal and
Opaque-2 Maize and Casein as Sole Protein Source
in the Clinical Treatment of Young Children
(After Bressani, 1975; Quicke, 1982)

Protein Source	N-Balance Index* (%)
Normal Maize	31.3
Opaque-2 Maize	72.5
Casein	80.0

* NBI, expressed as a percentage, is equivalent to
  biological value.

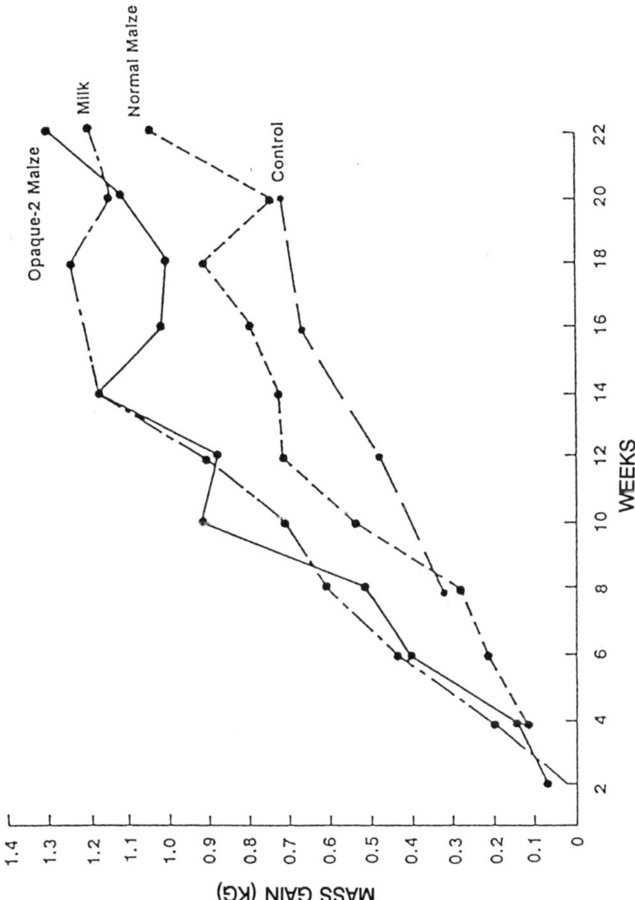

Fig.5. Response of 18- to 30- month old children from low income Indian families when fed opaque-2 maize, normal maize and milk as supplement to the home diet. (By kind permission of National Academy of Sciences, Washington.D.C., 1988, quoting results of Food and Nutrition Board, India, 1977).

apply to adults as shown by numerous authors (See a review by Quicke, 1982).

The dietary improvements resulting from HLM may seem particularly important among the poorer communities of our country and other countries where maize is still the dietary staple. But their general utilization should be of much wider advantage in providing a more balanced diet to humans of all ages and backgrounds. These considerable nutritional benefits are still awaiting widespread industrial exploitation and promotion among the consumers of maize in South Africa.

The International Maize and Wheat Improvement Centre (CIMMYT) is the only other research organization that has had comparable results in the development of highly acceptable, modified opaque-2 maize populations, or quality protein maize (QPM). QPM is advocated for the developing countries of Africa, Asia and America as a solution to widespread malnutrition (National Research Council, 1988). As the local breeding work has advanced beyond populations, namely, high yielding commercial QPM hybrids, a definite nutritional, agricultural and industrial potential is thereby presented for the promotion of HLM beyond our borders in local or international co-operative programs (Gevers, 1989).

<u>Animal feeds</u>. The use of HLM in monogastric animal feeds and its direct industrial exploitation offers the greatest immediate financial rewards. A summary of the important nutritional components of the protein of normal and HLM (HL2) is shown in Table 8. Using high lysine maize in pig feeding trials, it was shown that 22 % of fish meal normally used in pig diets could be saved due principally to the increased lysine content of HLM (Kemm et al 1977). Potentially, this saving can be converted to considerable economic advantage by the feed manufacturer and on-farm users of HLM.

TABLE 8

Protein and Amino acid content of air-dry
HL2 and Normal (N) Whole Grain Maize

Component	Percent of Sample		Percent of Protein	
	HL2	N	HL2	N
Protein	9.32	9.29	9.32	9.29
Lysine	0.36	0.26	3.90	2.74
Iso-leucine	0.30	0.30	3.24	3.22
Leucine	0.80	1.04	8.51	11.54
Tryptophan	0.13	0.09	1.51	1.05

* Source: Maize Board and University of Natal

Although until recently less clearly demonstrated (Gous & Gevers, 1982), the potential economic benefits of HLM in broiler nutrition are similarly summarised in Fig. 6 with the aid of whole grain protein data (Prof. R.M. Gous, personal communication). The advantage in terms of increased lysine and tryptophan have clear and quantifiable economic implications and otherwise confirm that HLM

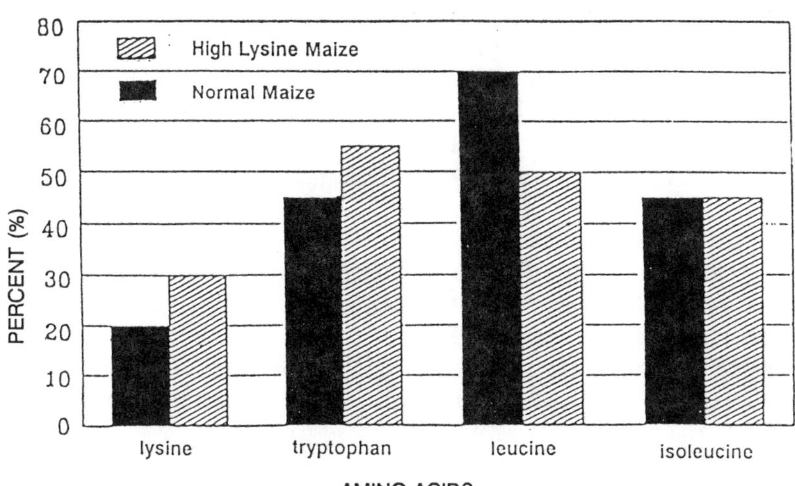

Fig.6. Concentration as percentage of requirement in broiler nutrition of four amino acids in maize.

has considerable potential value in the broiler industry. Estimates based on present (1989) production levels, feed requirements and costs of synthetic lysine, fish meal and other feed constituents, used to calculate least cost diets for both pigs and poultry (broilers and laying hens), indicate that the use of HLM instead of normal maize can effect an annual saving of R24.6 mil. in South Africa in the pig and poultry feed industry (Gevers, 1989)(Table 9). As synthetic lysine is at present being imported, its replacement with HLM would represent a considerable saving in foreign exchange.

<u>Soft versus hard kernel HLM</u>. Preliminary reports also suggest that the use of relatively soft whole grain HLM instead of hard-kernelled maize in cattle feedlots can bring about a saving of 3.7 to 10.5% of total diet costs, (Slabbert et al 1988, and personal communication). Ascribed in the first place to the softer kernel of the high lysine maize, this potential saving for an estimated 600 000 high grade carcasses produced in South Africa in 1988 amounted to between R5.9 mil. and R16.8 mil. These findings, for the first time, therefore also tentatively ascribe a potential value to HLM in ruminant feeding. Controlled experiments at basic metabolic level are in progress to evaluate more critically the contribution of HLM <u>per se</u> to these favourable nutritional results with ruminants. (Dr E.H. Kemm, personal communication).

The industrial potential of HLM in the pig and poultry feed industry therefore is considerable while unexpected possible advantages are also indicated for the ruminant. At a lower feed volume many of the possible industrial uses and benefits of HLM still remain to be exploited, but are no doubt potentially available to the manufacturer of feed for domestic animals, racing pigeons, fish and any other monogastric animal or organism which now utilizes normal maize in an industrially processed form.

TABLE 9
Saving (R) in Pig and Poultry Feed Costs when using HLM (1989/90)

Industry	Requirements (tons)		Lysine from maize(tons)			
	Feed	Lysine	HLM	Normal	Difference	Saving (R.mil.)
Pig	400 000	3400*	976*	576*	400	5.6 (11.8)**
Broiler	1 185 600	11 856	3 112	2 223	886	12.4 (7.5)
Layer	630 000	5 040	1 654	1 181	473	6.6 (9.4)
Totals	2 215 600	20 296	5 742	3 980	1 762	24.6 (8.7)

* Approximate   ** % Saving

<u>The economic and agricultural value of HLM</u>. While
the value of HLM in human food and nutrition can only
be indirectly assessed, its economic value in animal
feed production is for convenience usually directly
related to lysine saved although additional
nutritional benefits are also present. Estimates
based on present analytical data show that,
conservatively, every ton of HLM contains at least
one additional kg of lysine, which may be a realistic
minimum value of HLM to the producer, industry and
exporter. In these terms and at the cost of R14 000
per ton of synthetic lysine in 1989 this represents
a value of R14 per ton or 6.8% of the current
producer maize price of R207 per ton. Further, with
the prospects of exporting at least 5 million tons of
maize in 1989 (Maize Board, personal communication),
an added premium of R14 per ton would have meant an
additional earning of R70 mil. in foreign exchange.
The potential industrial value of HLM therefore
extends far beyond our borders. Together with all
its possible local financial benefits, it may be in
excess of R100 mil. annually (Table 10).

<u>The future of HLM</u>. The maize industry, as
represented in South Africa by the Maize Board, the

TABLE 10
Potential Total Value of HLM in South Africa(1989/90)

	Value (Mil. R.)
Pigs	5.6
Poultry	12.4
Layers	6.6
Ruminants	12.4*
Export	70.0
Total	107.0

* Approximate

National Maize Producers Organization (Nampo) and the producer, are the most important links in the chain of supply of maize to the consumer and industrial user of maize. If HLM is to achieve the nutritional and industrial potential that is indicated, then this supply chain should be developed and remain unbroken. Continuity of supply by the producer is dependent on a number of factors, the most important of which is that commercial HLM cultivars are competitive with conventional hybrids in terms of yield and other agronomic characteristics. Present data suggest that this is the case and that routine breeding progress may be expected in the future.

<u>Conclusions</u>. The major research and developmental work on HLM has been carried out successfully in the public sector with limited inputs and enterprise from the private sector. It is up to the industrial sector and the maize industry in South Africa to define future needs and strategies and to put pressure on the commercial seed companies to further develop and actively promote HLM if this promising new maize type is to reach its full agricultural, economic and industrial potential in South Africa.

ACKNOWLEDGEMENTS

Special acknowledgement and thanks for their contribution to the high lysine maize development program is due to Mr J.K. Lake, Chief Agricultural Research Technician, Grain Crops Research Institute and Professor G.V. Quicke, former Head, Biochemistry Department, University of Natal, Pietermaritzburg. Special thanks are due to Mr N.J. McNab for his sustained technical assistance and to Miss. S. McMillan for her secretarial inputs in compiling this report.

The biochemical analysis of local maize samples by Dr. E. Paiva of Embrapa, Sete Lagoas, Brazil, is gratefully acknowledged.

LITERATURE CITED

Alexander,D.E. 1966. Problems associated with breeding opaque-2 corn, and some proposed solutions. Proc. High Lysine Corn Conf. Corn Ind. Res. Foundation, Wash., D.C. 143-147.

Alvey, D.D. and Hamilton, J.R. 1967. Effect of genetic background on the lysine content of corn endosperm homozygous for the opaque-2 gene. Plant Breed. Abstr. 37: 1 (No. 572).

Bressani, R. 1975. Improving maize diets with amino acids and protein supplements. In High-Quality Protein Maize. Proc. CIMMYT-Purdue Symposium on Protein Quality in Maize, El Batan, Mexico, 1972. Stroudsburg, Pa, Dowden, Hutchinson and Ross.

Dennison, C. 1971. L-2,4-Diaminobutyric acid as a standard in the estimation of lysine in maize samples. J. Chromatogr. 63: 409-410.

Gevers, H.O. 1972. Breeding for improved quality in maize. Trans. roy. Soc. S. Afr. 40 (2): 81- 92.

Gevers, H.O. 1979. Advances in high lysine maize breeding. Proc. Third S. Afr. Maize Breeding Symp., Potchefstroom, 1978. Tech. Commun., Dept. Agric. Tech. Serv., Rep. S. Afr. No. 152: 51-57.

Gevers, H.O. 1982. High lysine maize - agronomic aspect. In "Rural studies in Kwazulu", (Eds. N. Bromberger & J.D. Lea). Proc. Symp., Univ. Natal, Pietermaritzburg. 14 Sept. 1981: 113-117.

Gevers, H.O. 1986. Review of progress in maize breeding with special reference to high lysine maize. Proc. F.X. Laubscher Memorial Symp., Potchefstroom 1986: 5-23.

Gevers, H.O. 1989. The industrial potential of high lysine maize. Proc. 10th SAAFOST Biennial Congress and Cereal Science Symp., Durban, Aug.1989, Vol II: 148-163.

Gevers, H.O. and Lake, J.K. 1990a. An agronomic comparison of opaque-2 maize breeding material and its near-isogenic counterpart. Proc. Ninth S. Afr. Maize breeding Symp., Pietermaritzburg, 1990. Tech. Commun. Dept. Agric. Developmt., Rep. S. Afr. (In print).

Gevers, H.O. 1990b. Progress in maize population improvement. Proc. Ninth S. Afr. Maize breeding Symp., Pietermaritzburg, 1990. Tech. Commun. Dept. Agric. Developmt., Rep. S. Afr. (In print).

Gous, R.M. and Gevers, H.O., 1982. An evaluation of high lysine maize in the diets of broilers. S. Afr. J. Anim. Sci. 12: 135-142.

Griffing, B. 1956. Concept of general and specific combining ability in relation to diallel crossing systems. Aust. J. Biol. Sci. 9.

Kemm, E.H., Gevers, H.O., Smith, G.A. and Ras, M.N. 1977. The use of South African bred opaque-2 maize in pig growth diets. S. Afr. J. Anim. Sci. 7: 127-131.

Maize Board. 1988. Yellow high lysine maize, Maize Board, P.O. Box 669, Pretoria.

Mertz, E.T., Bates, L.S. and Nelson, O.E. 1964. Mutant gene that changes protein composition and increases lysine content of maize endosperm. Science 145: 279-280.

National Research Council. 1988. Quality protein maize. National Academy Press, Wash., D.C.

Quicke, G.V. 1982. High lysine maize- nutritional aspect. In "Rural studies in Kwazulu", (Eds. N. Bromberger and J.D. Lea). Proc. Symp., Univ. Natal, Pietermaritzburg. 14 Sept. 1981: 119-129.

Slabbert, N., Bolt, M.J., Shelby, T and Campher, J.P. 1988. The utilization by beef steers of a soft high lysine and a hard normal maize cultivar in whole grain feedlot diets. Anim. Feed Sci. and Technol. 21: 31-41.

Van Twisk, P., Quicke, G.V. and Gevers, H.O. 1976. Physical properties and biological evaluation of high lysine maize. Cereal Chem. 53: 692-698.

Vasal, S.K., Villegas, E., Bjarnason, M., Gelaw, B. and Goertz, P. 1980. In "Improvement of quality traits of maize for grain and silage use".(Eds. W.G. Pollmer and R.H. Phipps) Martinus Nijhoff Publishers, 37-73.

Vorwerck, K., and Miecke, N. 1973. Möglichkeiten einer Bewertung der Verarbeitungseigenschaften von Mais. Die Mühle und Mischfuttertechnik 110:543.

Wallace, J.C., Lopes, M.A., Paiva, E. and Larkins, B.A. 1990. New methods for extraction and quantitation of zeins reveal a high content of Γ-zein in modified opaque-2 maize. Plant Physiol. 92: 191-196.

POTENTIAL ROLE OF QUALITY PROTEIN MAIZE
IN SUB-SAHARAN AFRICA*

Norman E. Borlaug
Department of Soils and Crops
Texas A&M University
College Station, TX

ACHIEVING A GREEN REVOLUTION; SORGHUM AND MAIZE
IN AFRICA

Newspaper stories and television programs: --
depicting the hunger, starvation and human misery in
African nations over the past six years -- have
resulted in the initiation of an avalanche of
assistance programs by European, Canadian, Australian
and American governments, as well as by international
and private organizations. As a result, there are
currently many hundreds of foreign agricultural
technical assistance and development projects in
operation in Africa, Asia and Latin America, designed
to assist the developing nations to improve their
agriculture, food production and standard of living.
Few of them have action programs to correct these
ills. Unfortunately, learned academic reports do not
necessarily equate to more food production. Most of
these projects are relatively ineffective, despite the
huge sums of moneys that are being poured into them.
As I see it, there are four primary reasons why most
foreign agricultural technical assistance programs are
relatively ineffective, if not doomed to failure.

---

* Read at the Annual Meeting of the American
   Association of Cereal Chemists, Dallas, Texas,
   October 17, 1990.

These are:
1.	Program plans and budgeting commitments are of
	too short a duration.
2.	Lack of continuity in assigning expatriate
	scientists to foreign assistance programs.
3.	Inadequate coordination of research effort across
	scientific disciplines, and lack of research and
	extension program linkages with economists and
	political leaders, who establish economic policy.
4.	Inadequate provisions for advanced degree
	training for young scientists of the host
	country, so that they will be prepared to carry
	on the program effectively when the foreign
	assistance program is terminated.
Now a brief statement about each of these factors
which limit the effectiveness of the technical
assistance programs.

In most developing nations, as is the case in
most African nations at present, there is an extreme
shortage of trained scientists. Moreover, either
there are no research and extension organizations, or
they are very weak and ineffective. Under such
circumstances, there are no easy short-term fixes.
Consequently, a "donor" country that is considering
establishing a foreign technical agricultural
assistance program in a developing nation must commit
itself to supporting such a program for a minimum of
ten years, and preferably fifteen, if it is to produce
a lasting and positive impact.

To be effective a foreign technical assistance
program must be staffed with a small number of
outstanding, well-trained, well-motivated,
enthusiastic scientists who are capable of working
effectively and tenaciously with the scientists of the
host country. They must also be self-motivated to
learn as much as possible of the history and culture
of the host country and to develop at least a working
knowledge of its principal language. By so doing, one
establishes credibility and gains the confidence of
the technicians, scientists, farmers and government
leaders. Expatriate-scientists must have good health
and outstanding physical stamina, which will enable
them to work effectively side-by-side with the young

scientists of the host country under difficult "back-country" conditions. The expatriate scientists must be "hands-on" working scientists, not consultants, if the program is to be successful. And above all, they should have leadership ability and personalities capable, by example, of stimulating the host country's young scientific colleagues.

Continuity of scientific research personnel is essential if an effective program is to be developed. It takes at least three years of imaginative and dedicated experimentation for a scientist in each discipline (e.g., plant breeding, agronomy and soils, weed, disease, and insect control), to establish an effective order of priorities to attach and ameliorate the major factors that impose bottlenecks on yield. It takes another three to five years to develop, through an aggressive on-farm research program, an appropriate and well-tested production package to extend to farmers. Unfortunately, because of donor government policies in many, if not most, foreign assistance programs, scientists remain in a given post for only two to three years before being moved to a similar post in another country, often with a different culture and language, with different soil and climatic conditions, and often to work on another crop.

In many foreign assistance programs, many of the scientists are professors who have been on detached assignment from a university in their home country, after which they return home, to be followed, in turn, by another professor on leave for a similar period. Such lack of continuity of scientific personnel is self-defeating. An effective, reliable package of improved production technology cannot be put together without continuity of capable, dedicated research scientists in each of the scientific disciplines, working at both the experimental station and farm level. All too often, to make matters worse, much of the time of these expatriate scientists is spent in writing reports to satisfy bureaucrats of both the national and international breed or for self-advancement in their own organization. With all of these handicaps, is there any wonder why many

technical assistance programs are ineffective? The
most frequently missing component, required to make a
foreign agricultural assistance program effective in
increasing production, is the "integrator".

I have found from experience that, no matter how
excellent and spectacular the research done in one
scientific discipline, its application in isolation
will have little or no positive effect on crop
production. Consequently, I have been forced to
become an integrator across scientific disciplines, in
order to assemble a reliable package of production
technology capable of dramatically increasing yields,
when it is properly applied.

Let me be perfectly clear; we need many well-
trained imaginative, scientific specialists in all
disciplines that bear on agricultural production. In
the developed nations, it is not difficult to identify
many outstanding scientists that have these
qualifications. But specialists alone will not
suffice in developing nations, where most agricultural
scientists do not have rural backgrounds, and where
research and extension organizations are weak and in
their infancy. In the developing nations we also need
a few of those rare "green-fingered scientific
integrators", call them "scrambling scientific
quarterbacks" if you prefer, who are comfortable,
willing and capable of venturing across the gray areas
between scientific disciplines to integrate the
composite knowledge into a package of improved
technological practices and see that it is applied to
produce more food.

Since I was originally trained as a forester,
permit me to use the analogy of a forest to emphasize
my point. It is always more comfortable for most
scientists to confine their research efforts to
his/her own specialty - and thus stand comfortably in
the thoroughly familiar and well-understood shade of
the tree of one's own discipline - while ignoring the
shade of the other trees. But the forest is made up
of the shadows cast by many trees, of different sizes
and often also of a considerable number of different
species. Consequently in order to best manage the
forest for multiple use purposes (e.g., lumber,

pulpwood and firewood production, watershed
protection, erosion and flood control, wildlife
habitat and recreation) on a sustained yield basis, a
few scientists are needed who not only have
outstanding credentials and skills in their own
disciplines, but at the same time can integrate across
the shadows or scientific disciplines cast by all the
trees of different sizes, ages and species in the
forest. Only with such research integration can one
develop and fit together the different pieces of the
jigsaw puzzle of production, and by so doing, assemble
a package of appropriate technology capable of
increasing and sustaining the overall productivity of
the forest systems to meet human needs.

So it is, also, with agriculture and food
production. The "green-fingered" scientific
integrator must assume the responsibility for
overseeing the evaluation of the new package of
technology on many farms, in order to establish that
it has the potential to increase yields by at least 50
to 100 percent, with acceptable levels of risk, over
that of traditional methods. While this is being
done, he must gain credibility and establish himself
as a concerned, trustworthy collaborator with
scientists, technicians and farmers of the host
country. Finally, at the opportune time (which is
difficult to determine, for in a large part it depends
on intuitiveness developed by intimate contact with
the program, culture and country), he must carefully
and skillfully wander into the swampy, "no-mans land"
of economic planners and political leaders who have
the responsibility and power to control economic
policies. They must be informed and convinced of the
potential increase in yield and production that is
possible if enlightened economic policies are embraced
which will stimulate widespread adoption of the new
technology by farmers, and hence, result in large
increases in food production.

THE SASAKAWA GLOBAL-2000 PROGRAM IN SUB-SAHARA AFRICA

Before trying to explain what the Sasakawa Global-2000 program is attempting to do through the introduction and production of high yielding Quality Protein Maize varieties and hybrids to improve nutrition in Ghana, it seems desirable to focus briefly on the current status of the food production and population issues of this vast and diverse region.

Sub-Saharan Africa is made up of 45 independent nations, with a population of 517 million, increasing at the rate of 3.0% per annum, which means it's population is increasing by 15.5 million annually and will double in 23 years, assuming the present growth rate continues. The vast majority of the population, varying from 55 to 90% in different countries, is engaged in agriculture and animal husbandry for their livelihood.

Sub-Saharan Africa constitutes a vast and diverse geographic region ranging across latitudes from approximately 20°N to 35°S. Ecologically it varies from desert to semi-desert in the north gradually changing toward the south to Savannah and then to tropical rain forest in the equatorial belt; beyond, to the south, it again changes and grades into Savannah and further to the south-southwest to semi-desert and desert, as in the Kalahari and Namibia. Its agriculture varies greatly depending upon the ecological zone. It goes from nomadic animal husbandry in the semi-arid regions of the Sahel, through slash and burn migratory agriculture, to "permanently" cultivated land in the more favorable ecological zones. Excluding the Republic of South Africa, most of the farms are small, subsistence family units employing only hand tools (hoe and machete) or animal powered equipment, and using little or no modern production inputs (e.g., chemical fertilizer, plant protection chemicals).

Thirty years ago African countries south of the Sahara were self-sufficient in the production of basic foods and exporters of specialty cash crops (e.g., coffee, tea, cocoa). However, over the past twenty years Africas' food production has grown at only half

of the population growth rate. The result is that
many of these countries are now spending much of their
scarce foreign exchange earnings on food imports
rather than on rural development projects.
Nonetheless, they are encountering worsening food
shortages on a recurring basis.

There are some intellectuals, including renowned
economic planners and scientists, who believe that
despite the huge land area potentially available for
cultivation, the problems of poor soils and adverse
climates impose severe restrictions on food production
with the technology now available. They believe that
it is unlikely, if not impossible, to trigger a
revolution in production of basic food within the next
two decades in Sub-Sahara, like the one that took
place in India, Pakistan, China, and Indonesia in the
1965-1980 period.

<u>Genesis of the Sasakawa Global-2000 African
Agricultural Program</u>. The famines that ravaged the
rural peoples of Sudan, Ethiopia, Uganda and some 20
other African countries south of the Sahara, following
the drought of 1983-84, greatly shocked the world.
One of the first to air-lift food aid to these people
was the noted philanthropist, Mr. Ryoichi Sasakawa,
Chairman of the Japanese Shipbuilding Industry
Foundation (JSIF). Although a long time, strong
supporter of United Nations International Disaster
Relief and World Health Programs, Sasakawa wanted to
attack the underlying causes of the African food
crisis and not just the symptoms. He was aware of the
role that the high-yielding Mexican wheat and IRRI
rice varieties, and the accompanying package of
improved agronomic management practices, combined with
enlightened government economic policy (which
stimulated widespread adoption of the new technology)
had played in averting widespread famine in India and
Pakistan in the late 1960's and early 1970's. He was
aware that this breakthrough in production had been
achieved during the time when some intellectuals and
international agricultural specialists said that
continuing hunger and worsening famines in these
countries was inevitable.

Mr. Sasakawa got in touch with me in late 1984 asking: "Can a 'Green Revolution' be triggered in sub-Saharan Africa?" I informed him that I knew absolutely nothing about African agriculture south of the Sahara and moreover, I was too old to learn at this late date in my life. His reply: "I'm 15 years older than you are - there is no time to waste - so let's start work on the project tomorrow." Thus, unwittingly and inadvertently, I became involved in another foreign technical agricultural assistance program - at this ripe old age.

For nearly two decades, food production, in most sub-Saharan countries has not kept pace with demand as explosive population growth and declining soil fertility have overwhelmed traditional agricultural systems. Despite the fact that 70-85% of the people in the majority of African countries are engaged in agriculture, most governments either have given agricultural and rural development a low priority or have pursued impractical, idealistic development goals.

Investments in production inputs, and their delivery, grain storage and marketing systems and in agricultural research, extension, and education have been woefully inadequate, and cheap food policies (costly subsidies) to appease the politically volatile urban dwellers have greatly distorted production incentive for farmers. Tribal ownership rather than individual ownership of the land also is a disincentive to the adoption of new technology.

<u>Sub-Sahara Africa Workshop</u>. As the first step toward developing pertinent information and knowledge on African agriculture, a workshop was held in Geneva, Switzerland on July 8 and 9, 1985. This workshop brought together 35 people from North America, Europe, Asia and Africa, who had interest and experience in African problems. It included agricultural scientists, economists, demographers, sociologists, economic policy specialists and religious leaders. It also included the Directors of the International Maize and Wheat Improvement Center (CIMMYT) and the International Crop Research Institute for the Semi-

Arid Tropics (ICRISAT), whose organizations had been assisting many of the sub-Saharan national programs with maize and sorghum research.

At this workshop the discussions centered on the information available/or needed to launch successful production programs, in two or more African countries, on maize and sorghum, the two most important cereal crops south of the Sahara. The development of the technology and its transfer from research stations to farmers fields, and changes in economic policies that led to the revolution in wheat production in India and Pakistan, were discussed in detail. It was hoped that these experiences might be relevant and helpful to guide in formulating a plan to launch production programs in maize and sorghum in a number of sub-Saharan African countries.

Mr. S. Subramaniam, who was Minister of Agriculture and Food of the Government of India during the food crisis of 1965-66, explained the importance of formulating economic policies that would encourage the adoption of new high yield technology. He spoke of the resistance he had encountered from some of his senior wheat scientists and economists when he imported the 18,000 tons of Mexican wheat seed and changed economic policies on grain prices at harvest, availability of production inputs (e.g., fertilizer) and credit. These changes in policy, he stated, made it possible for millions of small farmers to adopt the use of the package of improved production technology, which in turn revolutionized production and resulted in a four-fold increase in wheat production over a period of 20 years.

By the end of the workshop, it was clearly evident that in a number of African countries considerable research data and improved germplasm for both maize and sorghum were available with the potential for increasing yield. Unfortunately, they were lying largely unused in government experiment stations and laboratories. Moreover, there was additional available pertinent unused information in the three International Agricultural Research Centers (CIMMYT, ICRISAT and IITA) that had been developed in collaboration with a number of national African

programs. This information, too, was not being
transferred to farms in order to produce more food.

I informed Mr. Sasakawa (as well as Ex-President
Jimmy Carter and Mr. Hasan Abiede, President of the
Bank of Commerce and Credit International (BCCI), who
were both interested in African development and had
been active participants in the Workshop, that I
believed a package of improved agronomic management
practices potentially capable of substantially
increasing yield could be assembled from the available
research data and germplasm. I indicated that if a
program was to be established, it should be an
extension program (not another research program). It
should be designed to transfer and demonstrate
aggressively, on farmers' fields, the value of
improved production management technology. Moreover,
it was stated that such a program, if initiated, had
to become involved indirectly with government policy
issues:
1.   on the availability of production inputs e.g.
      seed of improved varieties, and fertilizer,
2.   with credit for purchases of inputs and
3.   with remunerative grain prices at harvest.

<u>Sasakawa Global-2000</u>. Mr. Sasakawa decided to sponsor
modest sized extension programs for five years in two
countries. In January 1986, Mr. Sasakawa, Ex-
President Carter, Dr. Leslie Swindale, Director of
ICRISAT, and I visited five African countries to
discuss the proposed program with Heads of
Governments. As a result it was decided to establish
Sasakawa Global-2000 agricultural extension programs
in Ghana and Sudan. Programs were initiated in 1985
involving maize and sorghum in Ghana and sorghum and
wheat in Sudan. Two expatriae experienced
agricultural scientists were assigned to each country.
They were to work directly with the national extension
service of the host country in training extension
agents in the proper use of the package of improved
management practices which would demonstrate the
potential value of this technology, in one acre plots,
on thousands of small farms.

Similar programs were initiated in the Fall of
1986 in Zambia under BCCI sponsorship, and in
Tanzania, Benin and Togo under Sasakawa sponsorship,
in 1988, 1989, and 1990, respectively.

<u>Program Philosophy and Distinguishing Features</u>. While
one blueprint does not fit all six agricultural
projects in Ghana, Sudan, Zambia, Tanzania, Benin and
Togo, all share common philosophical and programmatic
elements. First, all of the projects are concerned
with improving productivity in staple food crops grown
by small-scale men and women farmers (1 to 3 acres).
Second, we selected countries where we knew sufficient
research products and information had been generated
which were appropriate for small scale producers, but
which were not reaching them for various reasons.
Third, each of the projects is quite small, both in
terms of staff and financial resources. One to three
internationally recruited scientists are assigned to
each country project, except Benin and Togo, which are
handled by the Ghanian staff. These scientists work
with national counterpart staff in national extension
and research organizations.
     All of the Global-2000 project staff are highly
trained, experienced agricultural researchers. This
personnel staffing was intentional and designed to
give cooperating extension organizations more
creditability with research organizations. We believe
that technology generation and technology transfer
process must be an integrated activity involving
researchers, extension agents, and farmers, as was
designed and implemented to trigger the revolution in
wheat production in India and Pakistan in the 1960's.
In most developing countries, information flows from
research to extension are tenuous, and feedback
mechanisms from extension to research are almost non-
existent. We believe that it is time for extension to
be accorded more prestige by agricultural research
institutions.
     The Project staff emphasize overcoming the most
pressing constraints in staple food production first,
before moving on the other agricultural problems
and/or production opportunities. We are not waiting

for the "perfect" technology before trying to help
small-scale farmers. Rather, we believe that we can
help the farmer take the first step toward increased
productivity and an improvement in standard of living
with existing research products and information.

<u>Soil Fertility</u>. At present, soil infertility is the
most important factor limiting increased productivity
in all six countries. With little organic manure
available, the projects recommend moderate use of
chemical fertilizers to restore soil fertility, in
conjunction with improved varieties and more optimum
agronomic practices, so that farmers obtain greater
returns to their investments.

Our field testing and demonstration programs
differ from many other technology transfer efforts in
several important ways. The first difference is the
size of our demonstration plots, which are much closer
to commercial scale. Our demonstration plots are 0.4
ha (1 acre), thus providing the farmer with a
realistic test of the recommended technology and also
offering a sizeable and immediate economic benefit.
The field demonstration plots are called production
test plots (PTPs) and are designed to let farmers
evaluate the improved technology on their own fields
and to train extension workers and farmers in the
recommended crop management procedures.

The second difference is in terms of empowerment.
We go beyond just providing information. We are
prepared to provide the cooperating farmers with the
inputs, on credit, needed to grow the demonstration
plots, recovering these short-term production loans,
made to cooperating farmers, at time of harvest.

The third difference involves the psychology of
change. Our field testing program begins on a small
scale and, if the results prove promising, is then
expanded rapidly to a large scale to build widespread
"grass roots" pressure on political leaders to get
agriculture moving. By 1989, only three years after
our projects were conceived, several hundred thousand
men and women farmers can be counted as direct alumni
of the SG 2000 and BCCI-G2000 field demonstration
programs.

<u>Technology Transfer</u>. The final difference between our
approach to technology transfer and many other
development programs is the activist stance that we
take toward changing agricultural policy. We do not
accept status quo. We are committed to influencing
government investment decisions and to increasing the
amount of capital flowing into agriculture and rural
development. We seek to accelerate institutional and
infrastructural development. We believe that poorly
functioning technology delivery systems are a far more
serious constraint today to increased food production
in sub-Saharan Africa than the lack of improved
technology, per se. Consequently, we are actively
involved in lobbying political leaders to develop
effective input supply and price strategies
(fertilizer, improved seed, credit, and grain prices)
needed to assist small-scale producers.

We try to keep small-scale farmers and their
improved economic welfare at the forefront of our
thinking and actions. Our task is clear: to make
small-scale producers richer, more knowledgeable, and
more in control of their economic destinies. We know
that improved agricultural technology alone cannot
solve all of the social ills plaguing low-income
countries. We are also aware that the technological
changes we advocate will invariably create some
"losers" but we are convinced that there will be many
more "winners"; and with exploding population growth
and static food production, worsening social unrest
and political instability wait in the wings. It is
time for less talk and fewer learned publications and
more action programs to move improved production
technology to farmers' fields on the food production
fronts in Africa!

<u>The Sub-Saharan Countries Have the Natural Resource</u>
<u>Base to Increase Their Food Production</u>. Following the
widespread drought of 1983-84, which resulted in the
famine in Ethiopia, Sudan and Uganda and widespread
food shortage and hunger in more than twenty other
countries, a number of internationally recognized
agricultural development scientists claimed that these
countries, because of poor soil and adverse climatic

factors, will not be able to generate an agricultural
revolution like the one generated in Asia (India,
Pakistan and China) during the late 1960's and 1970's.
     The results of the Sasakawa Global-2000 maize and
sorghum demonstration program in five sub-Saharan
countries over the past five years, indicates that
this is not necessarily the case but, rather, that
there is, indeed, a large unexploited food production
potential. Over the past five years more than 70,000
one-acre plots of maize and sorghum have been grown
under "supervised" improved management practices.
And, perhaps, three times as many additional farmers
are using parts of the improved practices. The yield
increases, and the enthusiasm of the small traditional
farmer clearly indicate the large unexploited
production potential.
     With very few exceptions, the demonstration plots
have yielded double, triple and occasionally,
quadruple that of the farmer's plot employing
traditional methods. Moreover, often when grain yield
doubled, farmer's income increased two and a half to
threefold (depending on the country's economic
policies - e.g. prices of fertilizer, seed and price
of grain at time of harvest). It has become clear
that the very small African subsistence maize and
sorghum farmer (1 to 2 acres, equipped only with a hoe
and machete) is every bit as cognizant of the value of
the new technology - as were his counterparts, the
small Indian and Pakistani farmers, when they saw the
demonstrated value of the improved wheat technology
twenty-five years ago. This forced their governments
to change economic policy which in turn permitted the
small farmer to adopt the new improved production
technology. Currently, in the aforementioned African
countries, the enthusiasm and clamor to participate in
the use of new technology by tens of thousands of
small farmers is pressuring their governments to
modify their policies on credit, availability of seed
of improved varieties and fertilizer, plus
remunerative price of grain at harvest.

<u>Value of Extension Workers</u>. The experiences of the
past 5 years of the Sasakawa Global-2000 African

programs clearly indicates that information and
improved plant materials are of no value unless they
are transferred to farmers' fields. There is a wide
"no man's land" between research programs and
extension programs. The employees of the extension
organization are inadequately trained to transfer the
improved technology developed by the national research
organization. Moreover, they are immobile and
miserably budgeted.

The International Agricultural Research
Institutes have unwittingly, indirectly, and
unintentionally contributed to the worsening of this
dilemma - by improving the lot of the research workers
while ignoring the intolerable condition of the
extension program workers.

The Sasakawa Global-2000 program during the last
5 years has clearly demonstrated that with short-term
"hands-on" training demonstrating how to handle the
new technology, combined with modest economic
assistance, including bicycles to make them mobile,
the extension program and its staff have clearly shown
that maize and sorghum yields and production can be
greatly increased. Farmers enthusiastically accept
the new technology and are already pressuring their
governments to make the changes in economic policy
that will permit them to augment their production and
income.

The Sasakawa Global-2000 Demonstration Program
has clearly shown that the transfer of improved
production management technology, based on the
research that had been already done but was lying
largely unused, can dramatically increase yield and
production. Before it can convert the potential
production into a national production reality it will
require changes in government policy on credit,
availability of production inputs and remunerative
prices at harvest. The progress in increasing
production that has been made in Ghana, for example,
has already resulted in the appearance of second
generation development problems, i.e. shortage of
storage or warehouse capacity, a lack of an effective
organization to handle procurement of grain by
government at time of harvest so as to stabilize

prices to protect the farmer, and to release the grain
procurements into the market later, when prices
increase exorbitantly and thereby protect the
consumer.

In conclusion, the experiences of Sasakawa
Global-2000 indicate there is a large unexploited food
production potential in the African countries south of
the Sahara. To convey the potential into reality will
require that governments must have the political will
to give higher priority to agriculture and rural
development by making large investments in rural
education, transportation and storage facilities,
credit, realistic price policies, and availability of
production inputs. These improvements in policy must
be made promptly to ward off worsening food shortages
while the population monster is being tamed through
effective programs of family planning based on general
education and economic incentives - which is a slow
process at best.

PROTEIN QUALITY MAIZE FOR GHANA

It has been more than 55 years since Dr. Cicely
Williams, working in Ghana (then the Gold Coast),
described and determined the cause of a severe protein
deficiency disease, which she named kwashiorkor. This
disease was especially common and severe among
children who were receiving most of their calories
from cassava and yams, with only a limited amount of
food intake coming from maize or sorghum. It has long
been known that the cereal grains are deficient in one
or more essential amino acids and therefore have
limitations for corrective use in balancing diets. In
the second and third decades of this century the
University of Illinois made a concerted effort to
overcome, in part, this nutritional defect by
selecting lines of maize with higher levels of
protein. Even though selections were developed that
had up to 26 percent of total protein they still had
the same nutritional limitations because of low levels
of lysine and tryptophan. No progress toward
overcoming this nutritional limitation in maize
occurred until 1964, when Mertz et al. reported

finding much higher levels of both lysine and tryptophan in a long known mutant strain of maize, Opaque-2 (O_2). Dr. Evangelina Villegas, in her presentation in this book, superbly covers the historical aspects of the research, both genetic and bio-chemical, that has been done on O_2 and other mutant genes that influence the amino acid composition of maize protein. She also explains the difficulties of using these genes effectively in breeding programs.

I would, however, like to make a comment concerning the immediate short-term reaction of cereal scientists, research organizations, and research administrators to the discovery of the impact of the opaque-2 gene on the amino acid composition of maize protein. There was enthusiasm everywhere, in universities, private seed companies, and government agencies as teams of chemists and plant breeders, entomologists and plant pathologists were formed to develop improved varieties of maize, barley, sorghum, and wheat with better nutritional properties. Interest intensified as the positive results from early feeding trials with rats were repeated with excellent results with swine and later with children in advanced severe stages of protein malnutrition.

Then the importance of the negative effects of the opaque-2 gene began to appear: 1) soft endosperm associated with the high lysine and high tryptophan, 2) grain more susceptible to fungal ear rot, 3) more severe insect damage both in the field and in storage and 4) lower grain yield. The problem of converting the high lysine, high tryptophan soft endosperm types to dent and flint grain types proved to be a formidable obstacle. Since most of the aforementioned organizations that had launched QPM breeding programs did not have adequate back-up analytical chemical laboratory support, they lacked reliable data to guide their geneticist in reconstituting the dent and flint endosperm types and repeatedly lost the opaque-2 gene during recurrent selection following backcrossing. As a result, most companies and organizations abandoned their QPM breeding programs in the 1970's.

<u>The Road to Hard Endosperm QPM</u>. During the late
1960's and throughout the 1970's when I was the leader
of the CIMMYT wheat research and production program I
had a unique opportunity to observe Drs. S. Vasal and
E. Villegas (with Dr. Ernest Sprague's support)
develop their QPM breeding strategy, their laboratory
analytical methods and field methodology that
eventually led to the development of high lysine and
high tryptophan (QPM) open pollinated varieties and
lines in both dent and flint grain types, that are
indistinguishable (except by chemical analysis) from
normal maize.

But the road to developing the QPM lines with
dent and flint endosperm, with acceptable disease and
insect resistance and with grain yield and agronomic
type as good as the best open pollinated varieties was
long, arduous, frustrating and strewn with pitfalls.
The conversion of the floury starch of the opaque-2
soft endosperm to dent and flint endosperm types is
accomplished by the accumulation of a large number of
modifier genes during each generation of recurrent
selection following backcrossing; it must be based on
accurate chemical analyses. I was fascinated by the
techniques developed for analyzing different parts of
an individual kernel, without destroying the embryo.
This was accomplished by using a micro-drill to remove
for analysis very small samples from small "islands",
"spots" or layers of starch of different physical
textures. Those individual kernels containing the
opaque-2 gene and a number of effective modifier genes
controlling starch texture in different areas of the
kernel, were later germinated, intercrossed, and
backcrossed and reselected repeatedly for dent/or
flint endosperm. Using appropriate samples and
analytical procedures, many thousands of samples were
analyzed to make sure the opaque-2 gene was not lost
while the endosperm was being reconstituted into dent
and flint type grain.

This is a superb example that illustrates
beautifully how interdisciplinary research was used to
solve a complex biological problem.

During the past four years the Sasakawa Global-
2000 program has been involved in multiplying under

contract with the best farmers, seed of the
recommended Ghanian maize varieties: Safita (Pool 16)
SR (Resistant to Streak Virus), Abarutia (Tuxpano) SR,
and Okomasa (La Posta) SR. The seed from these
increases has been used to sow the tens of thousands
of production training plots that have been planted on
farmers' land to demonstrate the value of the package
of improved management practices. All three of these
varieties were developed jointly in the Crop Research
Institute, CIMMYT and IITA, employing CIMMYT
germplasm. All three of the varieties have performed
well.

<u>Conclusions</u>. Now it seems to me, the time has come to
move aggressively and to evaluate the best QPM in the
CIMMYT program in Ghana. In January 1990 arrangements
were made for Sasakawa Global-2000 to assist the Crop
Research Institute, CIMMYT and IITA to evaluate the
vast and diverse amount of CIMMYT's QPM materials -
including open pollinated varieties, inbred lines and
experimental hybrids under several locations in Ghana.
Hopefully one or more QPM materials now available will
be found with a combination of characteristics that
make them competitive with the best normal open
pollinated varieties now in use commercially. I
personally am impatient and cannot live comfortably
until I see QPM become a commercial reality. I know
it is badly needed, and see no scientific reason why
it cannot be done soon. It has been 26 years since
the potential value of the opaque-2 gene was reported
by Mertz and colleagues. It is time, I believe, to
make a serious effort to put it into commercial use to
serve human needs.

QUALITY PROTEIN MAIZE DEVELOPMENT IN BRAZIL

Ricardo Magnavaca[1]
National Maize and Sorghum Research Center
Sete Lagoas MG, Brazil

<u>Introduction</u>. Maize ranks first among cereals being produced in Brazil. It is the most widely grown crop being planted in regions with a great environmental diversity: humid tropics, semi-arid tropics, acid savannas soil, and sub-tropical fertile soils. General indications for maize production in 1989 are summarized in Table 1.

The large number of producers, diversity in production environments and in technological levels make hard the process of technological development and extension. The utilization of improved maize seeds

Table 1. Maize general indications in Brazil, 1989.

Harvested area	(000 ha)	13,333
Production	(000 t)	25,000
Number of producers	(000)	3,000
Improved seeds use	(000 t) <u>1</u>/	137

Source: CFP, 1988
<u>1</u>/ ABRASEM, 1989

---

[1] Reprinted by permission of the author and publisher from Proceedings of the International Conference on Sorghum Nutritional Quality. 1991. Purdue Univ. Press, W. Lafayette, IN. pp. 75-82.

that last decade covered about 70% of the planted
area, decreased sharply to about 50% last year
(Abrasem, 1989).

This abnormal situation is due to high seed cost
and lack of credit, but may change rapidly. Most of
the improved seeds are hybrids (90%) and the remaining
10% are improved open-pollinated varieties. Farmers
that usually do not buy improved seeds are usually
sowing second generation hybrid seeds or advanced
generation improved varieties.

Approximately 68% of total maize production comes
from farms with less than 100 ha (Table 2).

Despite the fact that maize is becoming a cash
crop in the country, much of the production is still
obtained from farms with smaller landholdings.
Consumption at the farm level is relatively high (34%)
according to CFP (1987), most being used for animal
feeding. But the amount being utilized for human
nutrition in the form of traditional dishes is still
important.

QPM Breeding. Twenty-three open-pollinated quality
protein maize (QPM) varieties, selected for hard-
endosperm through modifiers of the opaque-2 gene, were
introduced from CIMMYT in 1983. These varieties
improved for high grain yield, hard-textured kernels
(flint or dent), earlyness, lodging resistance,
shorter plants, and good disease and insect resistance
brought back the attention of Brazilian breeders to
the improvement of protein quality. In the past,

Table 2. Farm size distribution among maize producers
          in Brazil.

Farm size (ha)	% Total area
< 10	15
10 - 50	41
50 - 100	12
> 100	32

Source: EMBRAPA, 1988

besides the fact that a few high yield soft opaque-2
hybrids were developed (Ribeiral, 1974), a frustration
happened due to unfavorable linkages associating the
opaque-2 genes to the grain's soft texture. Since
these 23 CIMMYT QPM varieties had almost overcome the
adverse linkage, the breeders of the National Maize
and Sorghum Research Center, at Sete Lagoas, started a
breeding program aimed to obtain improved QPM
varieties and hybrids adapted to different Brazilian
environments. We believe that unless quality protein
comes associated with high yield, in competition with
normal maize, the acceptance would be difficult. A
maize that is high in energy, with better protein
quality, and normal fiber and oil contents, becomes a
product of high nutritive value that can be produced
at low cost in the whole country.

The 23 QPM varieties were planted in the winter
of 1984 at Sete Lagoas, and seeds were increased by
sib-crossing within each population (minimum of 500
plants). In the summer of 1984 the 23 QPM varieties
and two normal checks (BR 105 and AG 301) were
evaluated at six Brazilian environments: Porto da
Folha SE (Summer), Porto da Folha SE (Winter), Sete
Lagoas MG, Goiania GO, Nova Prata RS and Cruz Alta RS.
A summary of results for the best performance
varieties, published by Magnavaca et al. (1988), is
presented in Table 3.

From the tested populations, the four with white
kernels presented in Table 3 performed well in
different environments with a tryptophan-lysine
percentage in protein from 60.5% to 100% above the
normal hybrid AG 301. In terms of yield they were
superior or equal to the well adapted normal
population BR 105. The highest yielding population
produced 4% less than the normal hybrid AG 301. The
Population 64 QPM, a white dent kernel type, was
chosen to be selected in our breeding program based on
its yield, shorter plant, lodging resistance and ear
disease resistance. After three cycles of half-sib
selection it was released with the name BR 451. Data
from the 3rd cycle of selection in BR 451 are
presented in Table 4 for the average of two locations
in 1988.

Table 3.   Mean ear weight, percentage of protein and
           tryptophan-lysine in protein of QPM
           varieties.

Material	Mean ear weight	Protein	Tryptophan	Lysine
	kg/ha	%	% protein	
Pop. 63 QPM	6559	9.85	0.86	3.86
Pop. 64 QPM	6510	10.28	0.80	3.62
Pop. 62 QPM	5697	11.16	0.70	3.21
La Posta QPM	6758	10.28	0.69	3.17
BR 105 (Normal)	6247	9.60	0.45	2.19
AG 301 (Normal)	7043	10.39	0.43	2.11

Source: Magnavaca et al. (1988)

Table 4.   BR 451 half-sib evaluation of 200 families
           (average of two locations).

Parameters	Half-sib families (100 per set)	
	Set 1	Set 2
Ear weight (kg/ha)	8027	8520
Ear weight range (kg/ha)	6701-9776	6698-10521
C.V. (%)	12.8	13.6
Genetic CV (%)	4.0	5.4
Additive variance (g/plant)2	193.91	399.93
Heritability (%)	23.03	41.55
Expected progress (%)	4.0	7.0

Source: EMBRAPA/CNPMS (1988) - Not published

On the plot basis evaluation, the BR 451 families
showed a high yield potential, and enough additive
variance to allow progress in the next cycles of
selection. The half-sib families selected for
recombination to obtain the improved variety presented

an average of 9296 kg/ha of ear weight and 0.80%
tryptophan in protein. It was given priority in
selecting for better husk cover.

The white kernel character of BR 451 was a very
important genetic marker at the farmer's level. More
than 99% of the maize being grown in Brazil is the
yellow kernel type. Since the opaque-2 gene is
recessive, any yellow grain contamination in BR 451
ears would be normal type and therefore easily
detected. Since the BR 451 release was aimed to be
grown by farms with smaller landholdings, this easy
detection of contamination is important in maintaining
the genetic purity of the variety, allowing the
utilization of advanced generations without genetic
erosion.

<u>QPM hybrids</u>. Considering QPM for general utilization
at large scale in Brazil it is necessary to select a
yellow kernel type. In that case hybrids are a must,
due to the better possibility of obtaining higher
yield and the necessity of buying seeds for each new
plantation. It will be difficult to distinguish
phenotypic differences between normal and QPM yellow
kernels unless chemical analysis is made.

In order to rapidly develop yellow kernel QPM
hybrids, family hybrids based on full-sibs derived
from CIMMYT QPM populations were selected in our
program. A 5 x 6 and 10 x 10 partial diallel of full-
sib families plus parents were tested in four
locations. The treatments were allocated to each
diallel according to cycle differences (Magnavaca et
al., 1989). The full-sib parents per se mean for
flint and dent types, full-sib crosses mean and
heterosis data for both partial diallels are presented
in Table 5, as mean ear weight (kg/ha) of four
locations.

The high average and specific heterosis detected
for non-inbred material (full-sibs) from
interpopulation crosses of yellow CIMMYT QPM
populations made possible the selection of high
yielding family hybrids as well as indicated superior
families that could be used for developing inbred
lines. These selected family hybrids are being used

Table 5.   Mean ear weight (kg/ha) of full-sibs per se
           and full-sib crosses, and heterosis data of
           two partial diallels tested in four
           locations.

	Partial diallel	
	5 x 6 (early)	10 x 10 (intermediate)
Full-sib per se mean (Flint)	4382	3894
Full-sib per se mean (Dent)	3610	4075
Full-sib per se general mean		3996 3985
Full-sib crosses mean	6026	5549
Average heterosis	2030 (33.7%)	1564 (28.2%)
High parent heterosis (best cross)	60.5%	58.8%

Source: Magnavaca et al. (1989)

for initial tests of yield potential and tester for
new inbred lines. Superior families were used for
inbred line development and a group of these lines
were tested in a top-cross at two locations in 1989.
A summary of the data for the five best top-crosses
compared with the tester and the average of four high
performance early commercial normal hybrids used as
check is reported in Table 6.

The tester used for crossing the yellow QPM
inbred line was a QPM family hybrid with intermediate
yield level and low tryptophan percentage in protein.
Nevertheless, a group of inbred lines in top-cross
with that tester demonstrated a high yield potential
and high percentage of tryptophan in protein. It is
reasonable to expect specific hybrid combinations
within that group of lines that can compete with the
normal check hybrids in yields and improve even more
the tryptophan percentage. Playing with that group of
lines for specific combinations of better looking hard

Table 6. Performance of yellow QPM inbred line top-
          crosses. Mean of two locations in 1989.

Treatment	Ear weight (kg/ha)	Protein (%)	Tryptophan (% protein)
Top-cross - 26	9694	8.52	0.89
Top-cross - 24	9539	10.06	0.78
Top-cross - 6	9262	8.97	0.78
Top-cross - 18	9245	8.98	0.79
Top-cross - 11	8943	9.41	0.82
Tester (QPM)	7788	10.06	0.42
Check hybrids mean (normal)	11739	9.41	0.36

Source: EMBRAPA/CNPMS (1989) - Not published

endosperm phenotypes will be possibly associated with
better yield and better protein quality.

In 1990 the program at CNPMS/EMBRAPA will have
the results of single-cross QPM hybrids with yellow
endosperm, with the possibility of producing three-way
or double-cross type hybrids. Inbred-line development
will continue and advances in molecular biology (plant
genetic engineering) will be incorporated in the
program. With a continued effort in breeding we
expect QPM hybrids to perform as well as normal
hybrids.

QPM/wheat flour mixture. Wheat production in Brazil
has increased in recent years but self-sufficiency has
not been attained. More recently the government has
cut the subsidy for wheat flour, increasing the price
drastically. One kg of subsidized wheat flour was
cheaper than 1 kg of maize meal. With a real price
for both products always comes the possibility of
using alternate flour for mixing with wheat. One of
the possibilities is maize meal, QPM flour being the
best option considering the human nutrition objective.

Due to the possible changes in breadmaking
quality of the different QPM genotypes a joint program

of research was established between EMBRAPA/CNPMS and EMBRAPA/CTAA (Agrindustrial Food Technology Center). Many aspects are being studied like the changes in QPM meal physical properties as reported in Table 7.

In addition to presenting a white color, BR 451 also has better meal extraction percentage when compared with other QPM varieties. The white color of maize flour is important when the product to be made is French bread. The QPM maize could be stored as whole grain and be ground as needed. The QPM whole grain has 3 to 5% oil, avoiding fat addition to dough when maize is added up to 20% in wheat/maize mixtures.

The water absorption of wheat/maize flour mixture comparing the flour of normal maize (BR 108) and a QPM variety (BR 451) is presented in Table 8. Both varieties are white kernel and originated from the same tuxpeño race.

Water absorption is very important for breadmaking. Higher water absorption improves loaf quality and the economy of the process. The addition of maize to wheat flour decreased the water absorption compared to wheat flour but it was much higher for the normal than the QPM variety. The comparison of wheat

Table 7.   QPM meal physical properties.

Variety	Endosperm Color	Color (Kent-Jones Unit) 1/	Meal extraction	PSD2/(%) <49 m
Popul. 63 QPM	White	5.9	65.79	90.0
BR 451 - QPM	White	5.6	68.85	92.2
Guanacaste 7940 QPM	White	6.5	63.04	89.0
Popul. 65 QPM	Yellow	11.7	68.11	91.9
Popul. 66 QPM	Yellow	11.4	68.21	88.2
Amar Cristalino QPM	Yellow	12.0	63.29	90.9

1/   Kent-Jones, 1967
2/   Particle Size Distribution (%)
Source: EMBRAPA/CTAA (1988) - Not published

flour and the mixture of wheat/QPM flour for different
QPM varieties is presented in Table 9.

The acceptability of a loaf by the general public
will depend on the loaf volume. A higher loaf volume
and specific volume is a quality needed for
breadmaking. The QPM variety used for mixing with
wheat affects the loaf volume and again the BR 451
variety was the best. The data on wheat/QPM flour
mixture show the variability among QPM varieties on
breadmaking quality. That is an aspect to be
exploited by breeders in their programs and also
emphasizes the need for such tests.

Table 8. Water absorption in normal (BR 108) and QPM
          (BR 451) wheat/maize flour mixture.

Sample	Water absorption (%)	Reduction in water absorption
Wheat flour 1 (WF1)	57.4 1/	-
WF1 (80%) + BR 451 (20%)	55.7	2.9
Wheat flour 2 (WF2)	59.6 2/	-
WF2 (80%) + BR 108 (20%)	53.1	10.9

Source: 1/ EMBRAPA/CTAA (1988) - Not published
           2/ Mazzari, 1983

Table 9. Breadmaking quality of pure wheat and
          wheat/maize mixtures.

Sample	Loaf weight (g)	Loaf volume (cm)	Specific loaf volume (cm/g)
Wheat flour (WF)	130	580	4.46
80% WF + 20% BR 451	132	500	3.78
80% WF + 20% La Posta QPM	133	480	3.60
80% WF + 20% Am. Bajio QPM	134	460	3.42

Source: EMBRAPA/CTAA (1988) - Not published

Diffusion strategy for BR 451. The usual way of releasing new varieties of hybrids in our research program has been to inform the extension service, seed industry and sometimes farms during field days. Since 1987 the CNPMS communication service is using a marketing strategy, improving drastically the diffusion and adoption. The mass communication system has been intensively used and the release itself becomes a scientific, economic, social and political event. Diffusion materials have been prepared with a more attractive layout.

For the release of the QPM variety BR 451 the diffusion target was the whole society (especially the consumers), and not only the seed industry and farmers. Most Brazilian farmers have never seen a white maize kernel and they could be prejudiced against it. In that case the releasing could fail in the beginning. If the whole society accepts the new technology, it will create a demand that will have to be met by farmers and the seed industry.

The message received by consumers stressed the white kernel of BR 451 as a new type of maize as well as the high quality protein and possibilities of utilization in new products in mixtures with wheat flour. The news came at the time that the government decided to cut the wheat subsidy.

The release day became a national event with the presence of authorities, politicians, representatives of industry and farm associations, and mainly the press. The event should become a national news and even the choice of the day of the week and hour is important. Usually a Friday morning is favorable because of the usual lack of news for Saturdays. Many short notes about the release were published by the press a few days before the event. The Agriculture Ministry was in the morning TV news having breakfast based on BR 451 products before traveling for the event. During the event extension materials were distributed, products based on QPM and small seed bags. A lunch based on typical maize products was prepared with the BR 451 meal for all participants.

The most important agriculture diffusion TV news, Globo Rural, with an average audience of 15 million

people has dedicated two reports to BR 451. The same
network has also dedicated an issue of the Globo Rural
magazine to a report about the new variety and has
distributed a small plastic bag with 16 seeds attached
to the cover. That magazine issue was sold to 200,000
people due to the expectations created by the seeds;
this represented a 58% increase in circulation
compared with normal issues. The TV network Globo
made a 30 second advertisement of the magazine in the
evenings for many days calling attention to the seed
bag.

All of the advertising generated the sending of
30,000 mail letters asking for a kit of information
and seeds. The communication group has kept all these
addresses for further studies related to adoption.

In 1989 the production of BR 451 basic seed was
19.3 tons. From this amount 9.4 tons were traded to
seed industries to be increased. In 1990 there will
be 1,400 tons of BR 451 commercial seeds to be sold.
This amount is sufficient to plant an area of 70,500
ha. Besides that, we have no data on the planted area
based on seed increase of BR 451 distributed
throughout the country in small seed bags by the Globo
Rural Magazine and kits of seeds mailed directly to
farms. That area may be large because the farmers
already had two years of seed increase.

The marketing strategy for the release of BR 451
has proved to be efficient and will open the door for
the release of QPM hybrids.

Future program development. The great acceptance of
the QPM variety BR 451 and the good perspective of
obtaining a yellow kernel QPM hybrid with high yield
has stimulated EMBRAPA to continue and increase the
research program on QPM in the following aspects:
1.   To continue the selection of BR 451 and to
     develop yellow kernel populations as a source for
     inbred line development. To select yellow kernel
     QPM hybrids with high yield and nutritive value.
2.   To develop studies at the molecular level for
     understanding the relationships between different
     classes of endosperm proteins and grain hardness,
     flour and protein quality. In that aspect it is

very important to maintain a high level of
interaction with US universities.
3. To evaluate the effect of QPM on human and animal
   nutrition among farms with smaller landholdings.
4. To study the wheat/QPM flour mixture with
   relation to breadmaking properties.
5. To develop new products made with a base of QPM
   for the food industry, facilitating their entry
   into the consumer market.
6. To stimulate human consumption of QPM in the form
   of typical products of the various regions.
7. To stimulate the substitution of normal maize by
   QPM for animal feeding especially for farmers
   that normally do not utilize protein supplements.
8. To introduce new QPM products and formulations in
   urban areas.

The EMBRAPA research group works in a country
where fighting malnutrition is an important issue.
QPM can be important not only for direct human
consumption but also will permit a substantial
increase in total supply of foods like meat, milk and
eggs. These objectives may be attained by using other
agricultural products, but maize is the most widely
grown crop in the country and the one that can offer
high nutritive value at lower cost if QPM is used.
Therefore, it is our obligation to offer that new
option to our society.

References

Associacão Brasileira dos Produtores de Sementes.
   Anuário ABRASEM 1989. Berasília, 1989: 132 p.
Empresa Brasileira de Pesquisa Agropecuária,
   Secretaria de Planejamento, Brasília, D.F. I.
   Plano Diretor da EMBRAPA: 1988-1992. Brasília,
   1988: 564 p.
Companhia de Financiamento da Producão (CFP) - Anuário
   Estatístico 1982-87 - Ministério da Agricultura,
   Brasília, julho de 1988: 148 p.
Kent-Jones, D.W. and Amo, A.J. Modern Cereal
   Chemistry. 6 ed. London, Taylor Garnett Evans, p.
   1967.

Magnavaca, R.; Paiva, E., Winkler, E.I., Carvalho,
    H.W.L.; Silva, Filho, M.C., Peixoto, M.J.V.V.D.
    Avaliacão de populacões de milho de alta qualidade
    proteíca. Pesq. agropec. bras., Brasília,
    23(11):1263-1268, nov. 1988.
Magnavaca, R., Oliveira, A.C., Morais, A.R., Gama,
    E.E.G., Santos, M.X. Family hybrid selection of
    quality protein maize. Maydica 34:63-71, 1989.
Mazzari, M.R. Fubá de milho branco e pré-gelatinizado
    por extrusão, em mistura com farinha de trigo,
    para producão de pães. II. Qualidade e avaliacão
    tecnológica dos pães obtidos. EMBRAPA-CTAA, Rio de
    Janeiro, Boletim de Pesquisa nr. 6, 15 p., 1983.
Ribeiral, U.C. Situacão do milho de alta lisina no
    Brasil. IN: Reunião de Literatura da Cultura do
    Milho no Estado de Minas Gerais. Belo Horizonte,
    PIPAEMG, 1974. p. 26-39.

QPM HYBRIDS FOR THE UNITED STATES

A.J. Bockholt and Lloyd Rooney
Soil & Crop Sciences
Texas A&M University
College Station, TX 77843-2474

INTRODUCTION

Although the discovery of opaque-2 corn's exceptional nutritive qualities was made by Mertz et.al. in 1963, it has been put to little use in the United States. Moreover, the new, hard-endosperm forms that are known as Quality Protein Maize (QPM) have not been used in the U.S.

Despite this lack of local recognition, however, QPM could become a noteworthy part of the North American corn industry. Any improvement in the fundamental nutritional value of corn is likely to create new, unique, and premium markets.

There is, for instance, distinct promise for breakfast cereals and snack foods, both of which are constantly criticized for their low nutritional value. For them, QPM might provide both a nutritious product and an important public relations breakthrough.

QPM varieties and lines with a yield potential equal to commercial cultivars have been released in Guatemala, Paraguay, Ecuador, Brazil, China, and Africa. However, no QPM's are being utilized in the U.S. at the present time.

<u>Heterotic Patterns</u>. The potential of exotic germplasm for providing new alleles in breeding programs is recognized by breeders. According to Hallauer (1978), exotic germplasm will receive greater attention in the U.S. in the future. In order to determine how to combine germplasm in a hybrid program, it is essential to examine the heterotic relationships among the materials available. It is in identifying the heterotic patterns that one maximizes heterosis in the interpopulation cross or in hybrids between lines derived from diverse parental sources.

To initiate the QPM program, 11 QPM populations were obtained from CIMMYT, (Vasal et. al. 1980), (Bjarnason 1990). These source populations are listed in Table 1.

Table 1
QPM Populations Obtained From CIMMYT

Population	Adaptation	Seed Color
61 Early Yellow Flint	Tropical	Yellow
62 White Flint	Tropical	White
63 Blanco Dentado-1	Tropical	White
64 Blanco Dentado-2	Tropical	White
65 Yellow Flint	Tropical	Yellow
66 Yellow Dent	Tropical	Yellow
67 Templado Blanco Cristalino	SubTrop.	White
68 Templado Blanco Dent.	SubTrop.	White
69 Templado Amarillo	SubTrop.	Yellow
70 Templado Amarillo Dent.	SubTrop.	Yellow
Trop X Temp.	-	Yellow

Inbred lines from the CIMMYT populations were crossed in a line X tester mating design with two modified single crosses, (B73 X B84) and (Mo17 X Va35) to form the testcrosses. The modified single cross (B73 X B84) represented the Reid Yellow Dent germplasm and the single cross (Mo17 X Va35) represented the Lancaster germplasm, the most widely used heterotic pattern in the U.S.

Results from the testcross data are presented in Table 2.

Table 2
Heterotic Relationship in Inbred Lines
Derived from the CIMMYT populations

Population	Testers	
	B73 X B84	Mo17 X Va35
Pop 61	-	-
Pop 63	+	+
Pop 65	+	+
Pop 66	+	+
Pop 67	-	-
Pop 68	+	-
Pop 69	+	-
Pop 70	-	-
Pool 29	-	+
Temp X Trop	+	-

1] + indicates excellent combining ability
- indicates poor combining ability

From the results of this study it was determined that Pop 66, 68, 69 and Temp X Trop populations have excellent combining ability and generally can be considered as belonging to the Lancaster heterotic group. Pool 29 had excellent combining ability with the Lancaster tester and was determined to belong to the Reid Yellow Dent heterotic group. Populations 63 and 65 combined exceedingly well with both testers and would belong to a different heterotic group than those prevalent in the U.S. Populations 61, 67, and 70 combined very poorly with both testers in this study. The inbreds from Populations 62 and 64 were not represented in this study as all selections had previously been discarded because of their poor agronomic traits, i.e., excessive lodging, poor plant types, disease susceptibility etc. The most promising inbred lines for the U.S. have been derived from Populations 63 and 65, (Kirubi 1991).

Inbred Line Development. Two basic techniques have been utilized in the development of inbred lines: (1) selection within the CIMMYT QPM populations and (2) conversion of standard U.S. inbreds to the QPM trait.

CIMMYT Selections. The standard pedigree selection process was followed in developing the inbred lines from the populations obtained from CIMMYT. During the early generations of inbreeding, the lines were planted in a high plant population and evaluated for standability and anthesis-silking interval. The lines also were evaluated for insect, disease resistance and heat and drought tolerance. No testcrossing was done in the early generations as we did not have any good QPM testers. In the S_4 generation we started testcrossing to normal modified single-cross testers in order to determine the heterotic patterns and yielding potential of the lines.

In earlier generations many lines were eliminated because of poor agronomic performance, sterility, etc. In fact the selections from Populations 62 and 64 were all eliminated because of their poor agronomic performance.

These lines were not evaluated for combining ability. It is highly probable that these Populations possess some excellent yield and grain characteristics, but they would have to be modified to improve their agronomics.

The QPM populations and selections selected at College Station did not indicate any photoperiod sensitivity; however, when the materials were tested on the Texas High Plains and the Corn Belt, many of the selections reacted to photoperiod. These lines were late flowering and grew very tall.

<u>Conversion of U.S. inbreds</u>. In the conversion of the U.S. lines, we used the opaque version of the line i.e $B73_{02/02}$. Unfortunately the opaque lines that we had in our program or were able to get from other public programs were very limited. None of the newer lines had been converted to opaque. The publically available opaque lines which we used in our program are listed in Table 4.

Table 4.
U.S. opaque lines converted to QPM.

Tx $29A_{02}$	$A632_{02}$
$T220_{02}$	$B73_{02}$
$T224_{02}$	$B84_{02}$
$T232_{02}$	$Mo17_{02}$
$A619_{02}$	$W64_{02}$

The opaque lines were crossed to the CIMMYT QPM populations and backcrossed either 1 or 2 times to the opaque. These populations were then put into the selfing program and the standard pedigree procedure was followed. Throughout the inbred selection procedure primary emphasis was placed on selecting for the hard vitreous genotypes along with selection for desirable agronomic types. Periodically the lines were evaluated for lysine and tryptophan to insure that these essential amino acids remained at the desirable levels.

<u>QPM Hybrids</u>. In 1988 we grew the first QPM experimental hybrids produced in the United States. These hybrids were composed of inbreds selected from the CIMMYT lines which we had selected and U.S. modified QPM lines. Data on the top yielding QPM hybrids and the normal food corn check hybrids are presented in Table 5.

Table 5
Results of White QPM hybrid yield test grown
at College Station in 1988.

Hybrid	Yield bus/acre
Exp 23 X Exp 191	125
Exp 62 X Exp 191	120
Exp 26 X Exp 70	118
Exp 2 X Exp 191	118
Conlee 113W (check)	117
Exp 9 X Exp 58	116
Exp 10 X Exp 58	115
Asgrow 405W (check)	108

Four of the experimental QPM hybrids exceeded the yield of our best yielding white food corn hybrid. Although the agronomic data are not shown in this table the QPM hybrids were equally as good as the food corn hybrids in all attributes measured. In general they flowered slightly earlier than the checks, had excellent standability, disease and insect resistance, and grain quality. Similar results were also obtained with the yellow endosperm QPM hybrids. These results were extremely encouraging and the decision was made to expand the QPM breeding efforts.

After 4 years of testing we will be proposing the release of a white QPM hybrid and yellow QPM hybrid for commercial production. The agronomic traits and grain quality for these hybrids are presented in Tables 6, 7, 8, and 9.

Table 6.
Comparison of agronomic traits of yellow QPM and food corn hybrids grown in 11 environments (1989-91)

Hybrids	Days to Silk	Lodging, %		Yield bus/ac
		Root	Stalk	
Tx802 X Tx814	74.9	4.3	3.4	129.0
Conlee 202	75.2	10.1	3.3	123.8

Table 7.
Grain quality data for yellow QPM and food corn hybrid from 10 environments (1989-91)

Hybrid	Test wt.	1000 Kern. wt./g	Den-sity	Peri-carp Rem.	Hard-ness % 1]
Tx802 X Tx814	61.1	264	1.31	3.4	41.0
Conlee 202	61.9	327	1.32	2.6	35.2

1] The lower the value the harder the grain.

The yellow QPM hybrid is composed of two inbreds Tx 802 and Tx814. Tx802 is a selection from the CIMMYT Pop 65 while Tx814 is a modified T224. The QPM hybrid was slightly earlier silking and had better standability then Conlee 202. The QPM hybrid on the averaged yielded 5.2 bushels more than the check hybrid (Table 6).

Among the quality traits the QPM hybrid had a slightly lower test weight, smaller seed size and approximately the same density as the check hybrid. The pericarp was slightly more difficult to remove and the hardness test indicated that the grain was slightly softer than the check hybrid (Table 7).

Although the quality traits for the QPM are slightly below those of the food corn check, they are far superior to the U.S. feed corns.

Table 8.
Comparison of agronomic traits of white QPM and white food corn grown in 8 environments (1989-91)

Hybrid	Days to Silk	Lodging, % Root	Lodging, % Stalk	Yield bus/acre
Tx 807 X Tx 811	77.1	6.6	1.2	138.8
Conlee 113W	77.7	10.8	5.3	137.1

Table 9.
Grain quality data for white QPM and white food
corn hybrid from 8 environments (1989-91)

Hybrid	Test wt.	1000 Ker. wt/g	Den-sity	Peri-carp rem.	Hard-ness 1]
Tx 807 X Tx 811	59.6	279	1.32	1.8	42.0
Conlee 117W	60.5	324	1.33	1.5	36.6

1] The lower the value the harder the grain.

The white QPM is a single-cross hybrid composed of the two inbreds, Tx 807 and Tx 811. TX 807 is a direct selection from the CIMMYT Pop 63 while Tx 811 is a modified Mo17.

The white QPM in comparison to the check was slightly earlier, had better standability and produced a slightly higher yield for the eight environments in which it was tested (Table 8).

Among the quality traits presented in Table 9, the QPM hybrid had a slightly lower test weight, and smaller seed which were slightly softer. However, the density and pericarp removal rating was very similar to the check hybrid.

Although the quality traits for the QPM's were slightly below those of the food corn check hybrids, they are far superior to the U.S. feed corns. The check hybrids are considered to be the best food corns in our area; thus, we concluded that the QPM's are very acceptable.

Some of our experimental QPM hybrids have also been tested in the central corn belt (Iowa, Illinois, and

Indiana). Several of the QPM hybrids were very competitive in yield to some of the best hybrids.

<u>Future Breeding Efforts</u>. In order to keep the QPM hybrids competitive it will be necessary to develop techniques to permit us to more quickly convert the newer normal inbreds to QPM. Using conventional breeding techniques, the inbreds first have to be converted to opaques and then introduce the modifiers for the QPM.

$$\text{Normal} \rightarrow {}_{0202} \rightarrow \text{QPM}.$$

Using the conventional techniques the QPM hybrids will always be lagging behind the best hybrids by 10-12 generations. At Texas A&M University we are presently investigating the possibility of incorporating some of the molecular genetic tools to facilitate the development of new QPM inbreds and populations. The ${}_{02}$ DNA probe has already been introduced into the program to facilitate the development of new ${}_{02}$ lines.

As the public becomes more aware of the advantages of the "value added" component of the QPM's, we anticipate that there will be a rapid increase in the production and utilization of these materials.

REFERENCES

Bjarnason, M. 1990. CIMMYT's Quality Protein Maize program: present status and future strategies. Proceedings of the 18th Congresso Nacional de Milho e Sorgo, Vitoria, Brasil.

Hallauer, A.R. 1978. Potential of exotic germplasm for maize improvement. p. 229-247 in D.B. Walden (ed), Maize Breeding and Genetics, John Wiley and Sons, New York.

Kirubi, Duncan. 1991. Determination of heterotic patterns of Quality Protein Maize (QPM) inbred lines. M.S. Thesis, Texas A&M University, 79 pgs.

Mertz, E.T., L.S. Bates and O.E. Nelson. 1964. Mutant gene that changes protein composition and lysine content of maize endosperm. Science 145: 279-280.

Vasal, S.K., E. Villegas, and C.Y. Tang. 1984. Recent advance in the development of quality protein maize germplasm at CIMMYT. p. 167-189 in panel proceedings series. Cereal grain-protein improvement. 6-10 Dec. 1982. IAEA, Vienna, Austria.

EVALUATING QUALITY PROTEIN MAIZE GENOTYPES BY REVERSED-PHASE HIGH-PERFORMANCE LIQUID CHROMATOGRAPHY[1]

Jerrold W. Paulis, Jerold A. Bietz,
Frederick C. Felker, and Terry C. Nelsen

National Center for Agricultural Utilization Research
Agricultural Research Service
U.S. Department of Agriculture[2]
Peoria, IL 61604

INTRODUCTION

The major storage proteins in maize are alcohol-soluble prolamins, consisting of zein and alcohol-soluble glutelins (ASG) (Paulis and Wall, 1971). ASG contains water-insoluble (wiASG) and water-soluble (wsASG) subunits (Paulis and Wall, 1977).

High-lysine opaque-2 (o2) maize contains less zein than normal genotypes, giving it a higher level

---

[1]Presented at the 75th Meeting of the American Association of Cereal Chemists, Dallas, Texas, October 14-18, 1990.
[2]The mention of firm names or trade products does not imply that they are endorsed or recommended over other firms or similar products not mentioned.

of lysine-containing proteins and making it more nutritious (Ruskin, 1988). o2 maize also has a soft, floury texture, however, so it may break more easily, be susceptible to ear rot, and give less grits upon dry-milling. To correct these problems, breeders wish to retain o2 genes for high lysine, while increasing endosperm hardness.

There is thus much interest in development of quality protein maize (QPM), or modified opaque-2 (mod. o2) genotypes (Ruskin, 1988). Maize endosperm texture also frequently relates to protein composition, which can be accurately and quantitatively determined by reversed-phase high-performance liquid chromatography (RP-HPLC) (Paulis and Bietz, 1986; Paulis et al., 1990, 1991) using ultraviolet absorbance detection at 210 nm (Scopes, 1974; Biemond et al., 1979). We therefore compared alcohol-soluble proteins of high-lysine mod. o2, high-lysine o2, high-lysine floury-2 (fl2), and normal maize genotypes by RP-HPLC to determine if amounts of specific alcohol-soluble glutelin or zein polypeptides could differentiate these genotypes (Paulis et al., 1991). We also examined developing and mature grains by scanning and transmission electron microscopy to see if protein differences relate to grain hardness as evident from endosperm ultrastructure.

MATERIALS AND METHODS

Maize Genotypes

Sixteen mod. o2 maize samples were supplied by R. J. Lambert (Univ. of IL), and six by E. M. Villegas [Centro International de Mejoramiento de Maize y Trigo (CIMMYT), Mexico]. The IL genotypes with common genetic background were the inbreds R802 o2 o2, R802 O2 O2 (normal), and an S4 R802 mod. o2 line. R802 was self-pollinated, and kernels were collected at various intervals after pollination. Endosperm sections were removed immediately and freeze-dried. The other mod. o2 IL lines were B73 mod. o2, TT mod. o2, Ms315 mod. o2, T/T mod. o2, E/I mod. o2, and Ia 36 mod. o2. CIMMYT genotypes Tuxpeño-1 (normal), Poza Rica (_o2 o2_), and Blanco Dentado-1 QPM (mod. _o2_, or Pop. 63) were of common genetic background. All CIMMYT QPM populations (63, 64, and 65) and QPM pools (17, 23 and 26) were derived from about 11 cycles of selection for modified endosperm.

Protein Extraction and RP-HPLC

Zein and ASG were simultaneously extracted from ground maize endosperms with 70% ethanol containing 0.5% (w/v) sodium acetate + 5% (v/v) β-mercaptoethanol or 0.2% dithiothreitol (DTT) (Fig. 1). Proteins were fractionated by RP-HPLC on a C_{18} column using an aqueous acetonitrile + 0.1% trifluoroacetic acid gradient. Eluting proteins were detected by absorbance at 210 nm, where peptide bonds absorb (Scopes, 1974; Biemond et al., 1979). Data were stored in a computer for subsequent integration

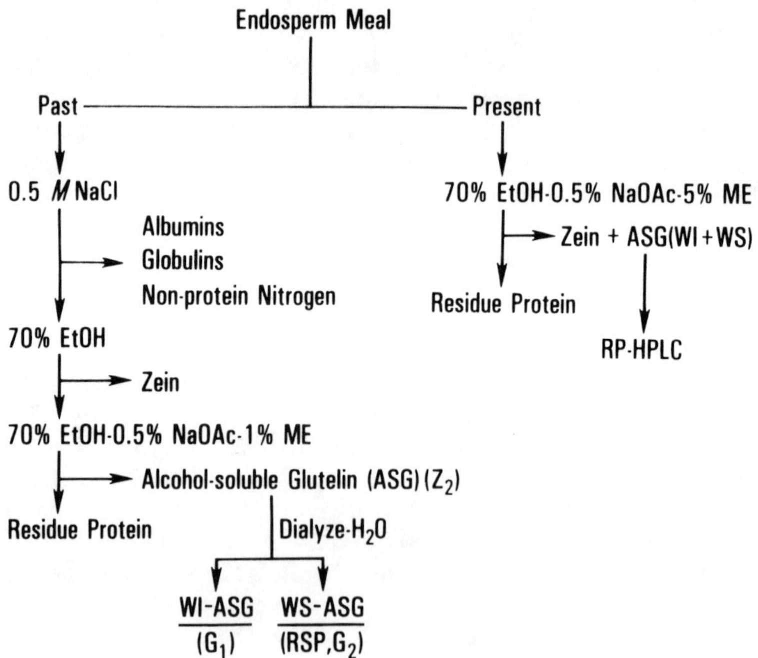

Fig. 1. Comparison of isolation schemes for alcohol-soluble maize proteins.

and statistical evaluation (Paulis and Bietz, 1986). For some studies, zeins were resolved further by modifying the gradient.

Microscopy

For transmission electron microscopy, subaleurone tissue was sampled from normal, _o2_, and mod. _o2_ R802 endosperms at 25 days after pollination. The tissue was fixed in glutaraldehyde, postfixed in OsO_4, dehydrated in acetone, and embedded in epoxy resin. Thin sections were stained with uranyl acetate and lead citrate. Mature grains of the same genotypes were viewed by scanning electron microscopy after transversely cracking the dorsal endosperm.

Protein Analyses

Nitrogen was determined on dried materials and extracts by a semi-micro Kjeldahl method. Samples of protein and endosperm meals (equivalent to 1-10 mg protein) were hydrolyzed by refluxing with 6 M HCl (2-10 ml per mg sample) for 24 hr. Liberated amino acids were determined on a Dionex D-300 amino acid analyzer using a physiological fluids analysis column with lithium buffers.

RESULTS AND DISCUSSION

Extraction and RP-HPLC of Alcohol-Soluble Proteins

Figure 1 compares our previous modified Osborne sequential extraction scheme for alcohol-soluble maize proteins to the method now used. In the previous scheme, zeins, extracted with 70% ethanol, represent 41% of total normal maize endosperm protein. Addition of a reducing agent to the aqueous alcohol extractant solubilizes additional alcohol-soluble glutelins (ASG), which can be separated by dialysis against water into wsASG and wiASG. Protein fractions similar to wiASG were designated G_1 by Landry and Moureaux (1970), and ASG has been called zein-2 (Z_2) by Sodek and Wilson

(1971).   Proteins comparable with wsASG have been
termed RSP (reduced soluble protein) and pH_3-G_2
(Wilson et al., 1981; Landry et al., 1983).

In contrast, the current method to extract
alcohol-soluble proteins uses only one extractant,
70% ethanol + 0.5% sodium acetate + 5% β-mercapto-
ethanol or 0.2% dithiothreitol (DTT), greatly
simplifying protein extraction.   Zein, wiASG and
wsASG are coextracted.   This extract may be directly
analyzed by RP-HPLC, revealing amounts and
relationships of these three protein fractions
(Paulis and Bietz, 1986).

Table I shows percentages of total alcohol
soluble proteins extracted from 6 different CIMMYT
mod. o2 (QPM) genotypes, compared with literature
values for normal and o2 endosperm types.   As in Oh43
normal and o2 endosperms (Misra et al., 1975),
alcohol-soluble proteins of these lines contained no
lysine, so lysine content and percentage of alcohol-
soluble proteins were inversely related.   Opaque-2
maize had the highest percentage of lysine (3.5%) and
the least alcohol-soluble protein (35%); CIMMYT mod.
o2 had an intermediate lysine content (2.7%) and 48%
alcohol-soluble protein; and normal maize endosperm
had the least lysine (1.6%) and the most alcohol-
soluble protein (65%).

TABLE I

Contents of Lysine and Alcohol-Soluble Proteins
in Maize Genotypes

Genotype	g Lysine/ 100 g Protein	Alcohol-Soluble Proteins (% of Total)
Normal[a]	1.6	65
Opaque-2 (o2)[a]	3.5	35
Modified o2[b]	2.7	48

[a] Misra et al. (1975).   Alcohol-soluble proteins
include zein and zein-like alcohol-soluble glutelin
subunits.
[b] Average of 6 CIMMYT QPM endosperm meals.

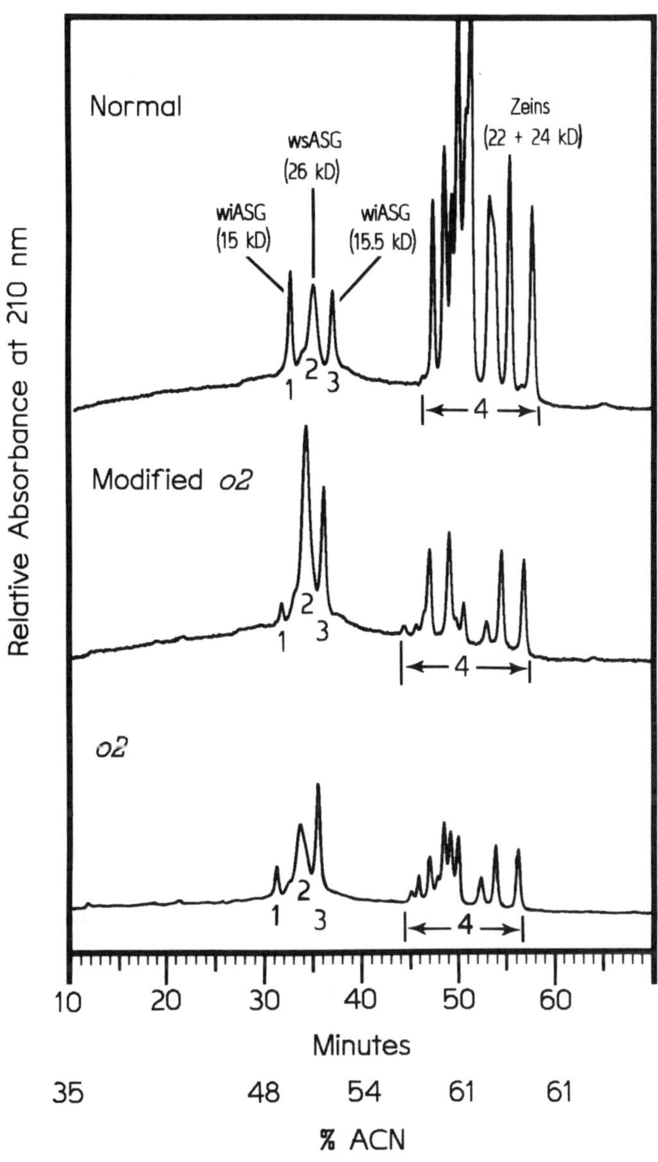

Fig. 2. RP-HPLC of prolamins from R802 normal, modified opaque-2 (o2), and o2 maize endosperm.

Amounts of alcohol-soluble protein extracted from R802 genotypes agree with percentages of total alcohol-soluble proteins from defatted and

non-defatted endosperms of the same strains
(Gentinetta et al., 1975; Robutti et al., 1974).
Alcohol-soluble protein contents of CIMMYT QPMs were
higher (45-49% of total protein) than in R802 mod.
o2, which might account for the increased content of
fraction 2 as compared to R802 mod. o2.

Figure 2 shows RP-HPLC patterns of alcohol-
soluble proteins from R802 normal, mod. o2 and o2
lines. Proteins in fractions 1-4 correspond in Mr
and amino acid composition, respectively, with mainly
(a) a 15 kD high-methionine wiASG subunit; (b) a 26
kD high-proline, high-histidine wsASG subunit [as
previously isolated by preparative RP-HPLC (Paulis
and Bietz, 1986)]; (c) a high-proline 15.5 kD wiASG
subunit; and (d) 22 + 24 kD zeins (plus a minor 10 kD
high-methionine wiASG subunit not shown),
respectively. The main components of fractions 1-3
were first described by Paulis and Wall (1977).
Mod. o2 R802 maize contained significantly more
early-eluting wsASG (fraction 2) and less zein than
did normal and o2 lines (Fig. 2). However, contents
of zein (fraction 4) were less in both mod. o2 and o2
maize endosperms.

Distribution of Alcohol-Soluble Proteins in Normal,
f12, o2, and Mod. o2 Genotypes

Relative amounts of the four alcohol-soluble
maize protein fractions in normal, o2, and mod. o2
genotypes are shown in Fig. 3. Protein distributions
in R802 and CIMMYT samples were similar. Relative
amounts of fractions 2 (wsASG) were significantly
higher in QPM (mod. o2) than in normal genotypes,
while amounts of zeins (fraction 4) were lower.
Average amounts from duplicate extractions for
RP-HPLC fractions 1-4 (as in Fig. 2) in IL mod. o2,
CIMMYT mod. o2 (QPM), o2, f12 and normal genotypes
are shown in Fig. 4. Relative alcohol-soluble
protein percentage data from other near-isogenic
normal, f12, and o2 genotypes (Paulis and Bietz,
1986) were averaged with three normals (Tuxpeño-1,
R802 O2 O2, B37 O2 O2), 3 o2 lines (Poza Rica o2 o2,
B37 o2 o2, R802 o2 o2) and one f12 maize genotype
(B37 f12 f12) of the same strain and having protein

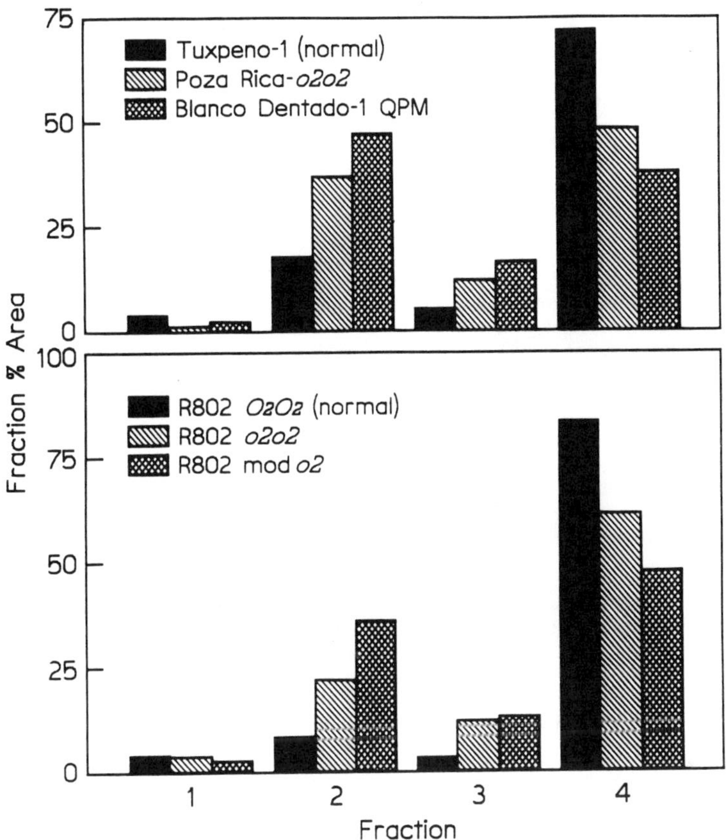

Fig. 3.   Comparison of RP-HPLC fractions 1-4 (see Fig. 1) from normal (Tuxpeño-1 and R802 <u>O2 O2</u>), <u>o2</u> (Poza Rica <u>o2 o2</u> and R802 <u>o2 o2</u>), and mod. <u>o2</u> (Blanco Dentado-1 QPM and R802 mod. <u>o2</u>) maize genotypes.

distribution in the range previously reported.   All genotypes showed good reproducibility upon RP-HPLC. Duncan's multiple range test was used to determine significant (p < .05) differences between genotypes of the four fractions.   All mod. <u>o2</u> lines had similar protein distributions and differed (p < .05) from normal genotypes and from <u>o2</u> and <u>f12</u> lines in that mod <u>o2</u> fraction 2 was considerably higher and fraction 4 was lower.   Amounts of fraction 2 were somewhat higher in CIMMYT mod. <u>o2</u> than in IL mod. <u>o2</u>

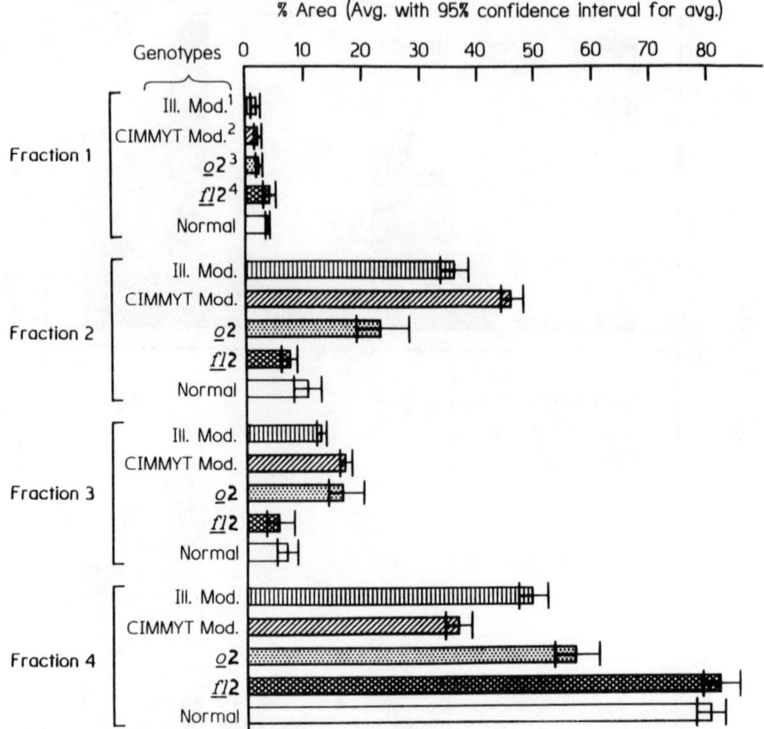

Fig. 4. Distribution of alcohol-soluble proteins in several genotypes of normal, opaque-2, and modified opaque-2 endosperms. (1) Ill. Mod = R802 mod. o2; (2) CIMMYT Mod = CIMMYT mod. o2; (3) opaque-2; and (4) fl2 = floury-2.

lines. This may be correlated to harder endosperm and increased amount of zein-like (G_1) or wsASG component in CIMMYT lines (Ortega and Bates, 1983).

Quantitative Variation in Protein Fractions
During Development

In an attempt to relate maturation of mod. o2 (QPM) endosperm to protein composition, we used RP-HPLC to determine relative amounts of alcohol-soluble protein fractions at different stages of kernel development of three R802 genotypes (Figs. 5-8). In normal endosperm at the earliest stage

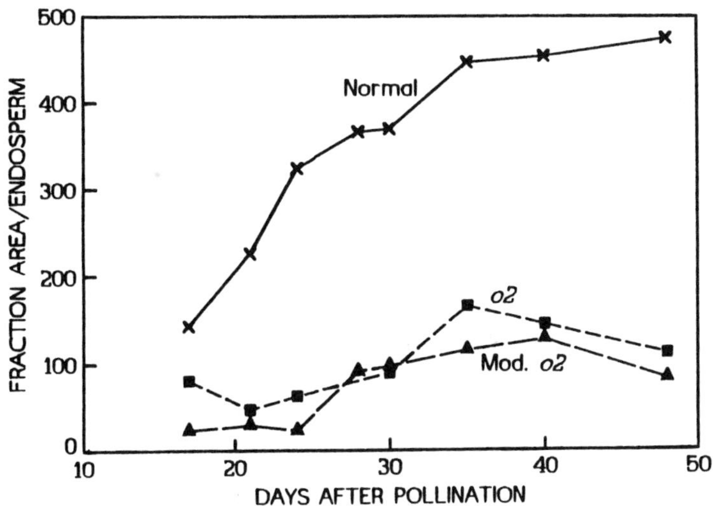

Fig. 5. Accumulation of RP-HPLC fraction 1 (15 kD high-methionine wiASG) in R802 maize endosperm during maturation.

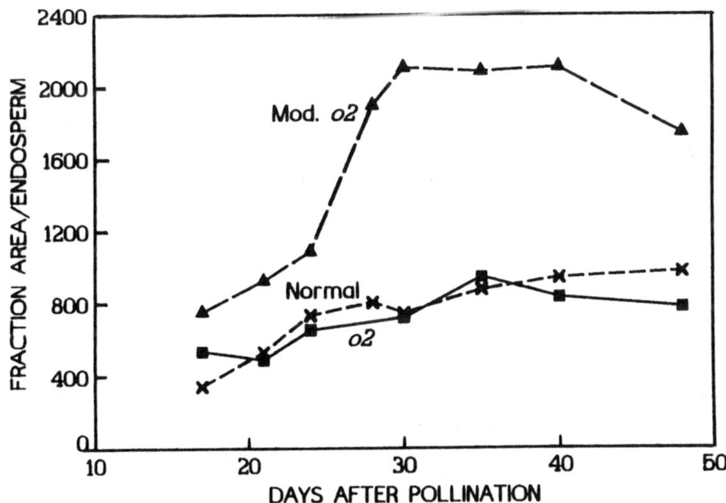

Fig. 6. Accumulation of RP-HPLC fraction 2 (26 kD high-proline wsASG) in R802 maize endosperm during maturation.

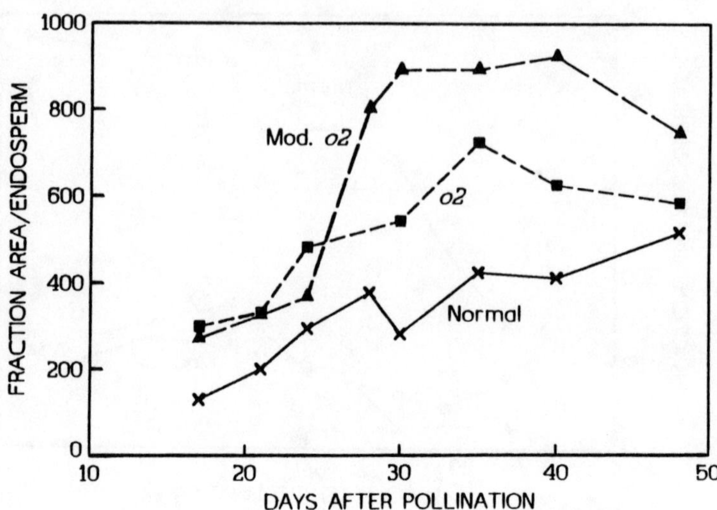

Fig. 7.   Accumulation of RP-HPLC fraction 3 (15.5 kD high-proline wiASG) in R802 maize endosperm during maturation.

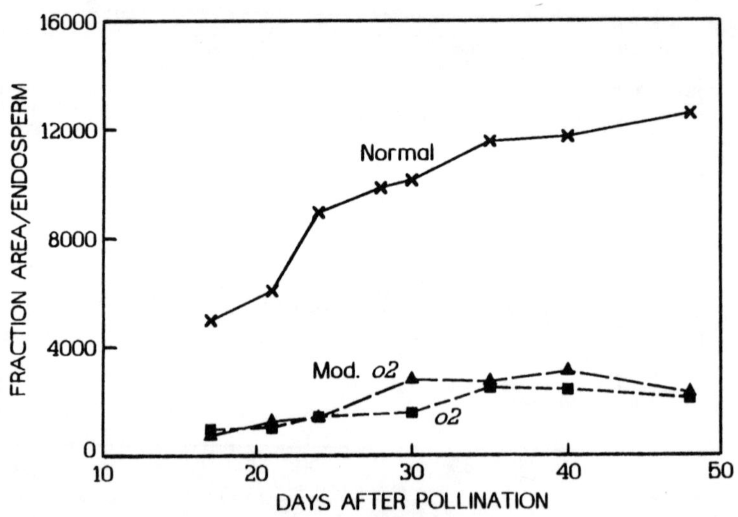

Fig. 8.   Accumulation of RP-HPLC fraction 4 (22 kD + 24 kD zeins) in R802 maize endosperm during maturation.

examined, the 15 kD wiASG was about three times more abundant than in _o2_ or mod. _o2_, and increased until 48 days after pollination (Fig. 5). The cDNA coding for this 15 kD wiASG was cloned and sequenced by Pedersen et al. (1986).

Figure 6 compares accumulation of high-proline wsASG (fraction 2) during development of normal, _o2_, and mod. _o2_ R802 maize endosperm. In mod. _o2_, this fraction increased from 800 area units at day 15 to more than 2000 units after day 30 during development, compared to an initial level of less than 600 units to a maximum of about 800 units after day 30 for normal and _o2_ endosperms. Das and Messing (1987) also reported different levels of this protein in W23 and BSSS53. The protein occurred along the periphery of protein bodies next to the outer membrane of this organelle (Ludevid et al., 1984). High-proline 15.5 kD wiASG (fraction 3) also increased most during development in mod. _o2_ maize (Fig. 7). Wallace et al. (1990a) also reported elevated γ-zein (fractions 2 and 3) in QPM.

Zeins (fraction 4) were much more abundant throughout development in normal maize endosperm than in _o2_ or mod. _o2_ genotypes (Fig. 8). Zeins in fraction 4 were resolved better by modifying gradient conditions (60 min linear gradient from 48 to 58% CH_3CN), as shown in Fig. 9. RP-HPLC results for zeins of _o2_ and mod. _o2_ genotypes were similar, though some significant qualitative differences did occur, suggesting differences in inhibition of 24 kD zein subunit synthesis (Soave et al., 1981). For example, peaks 8 and 12 were present and absent, respectively, only in mod. _o2_. Both _o2_ and mod. _o2_ genotypes contained considerably fewer components than did normal maize. Developmental stage especially affected peaks 10, 12, 13, 14, 15, 19, 20, and 22.

Microscopic Examination of Developing Kernels

Transmission electron microscopy of developing normal, _o2_, and mod. _o2_ R802 subaleurone revealed characteristic differences in size and appearance of protein bodies (Fig. 10). Although the distribution

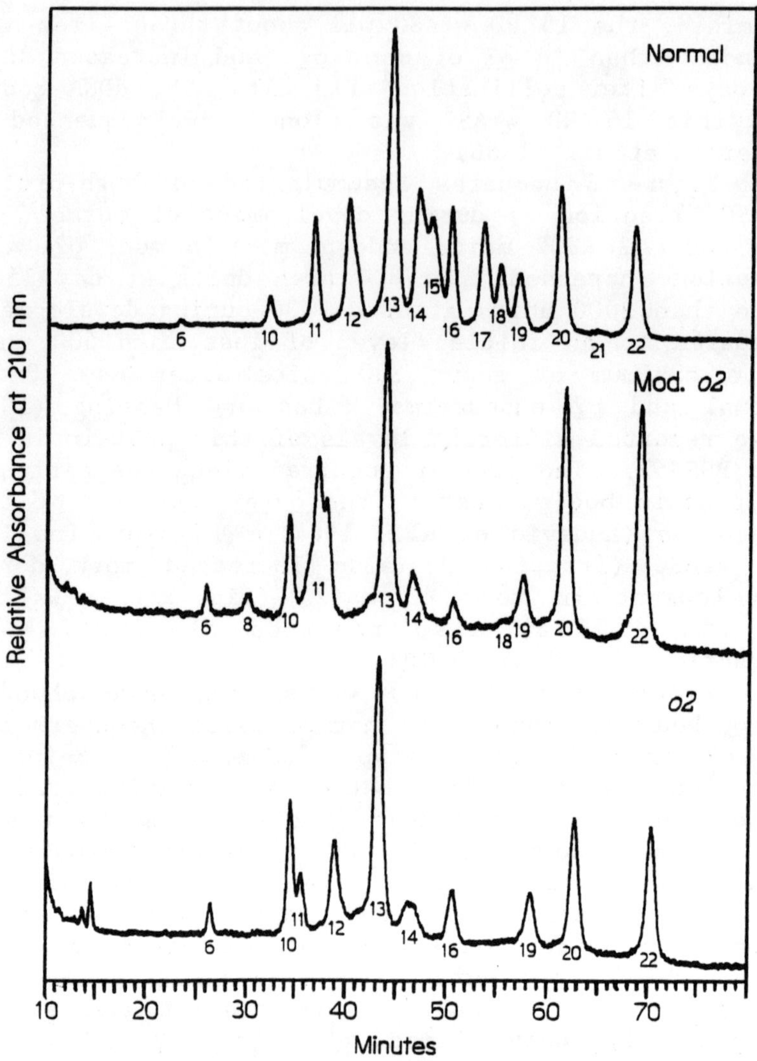

Fig. 9. RP-HPLC of zein proteins (fraction 4) from R802 maize endosperm genotypes. Two and three times more mod. o2 and o2 extracts, respectively, than normal extract were used to detect smaller peaks in these genotypes.

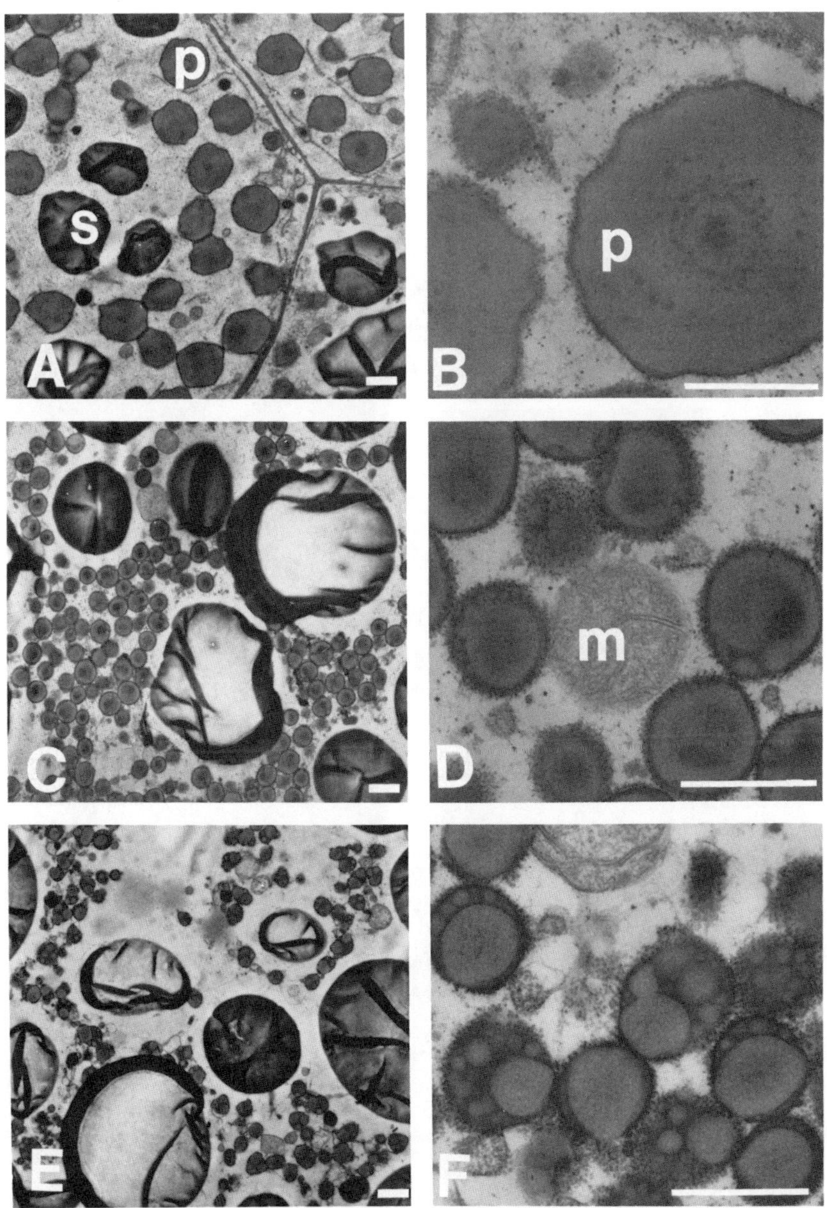

Fig. 10. Transmission electron micrographs of normal (A,B), _o2_ (C,D), and mod. _o2_ (E,F) R802 maize subaleurone tissue harvested 25 days after pollination. m, mitochondrion; p, protein body; s, starch grain. Bars represent 1 micron.

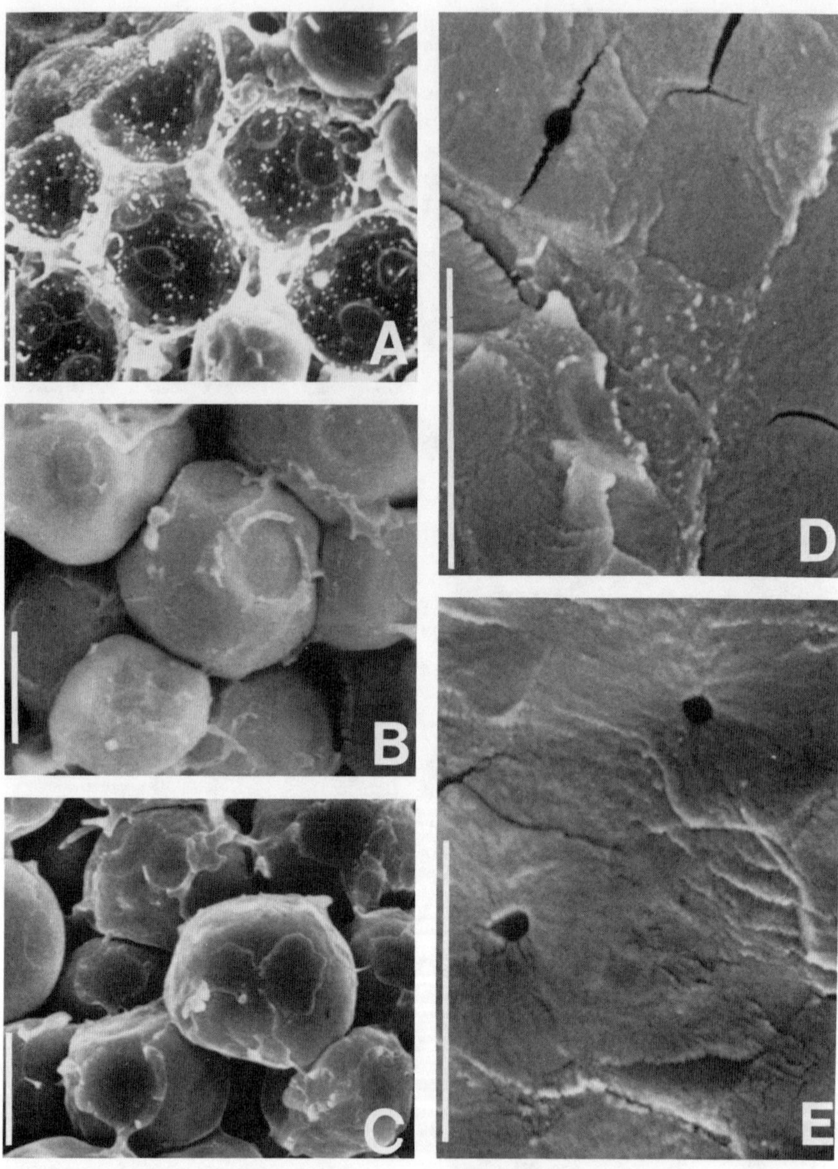

Fig. 11. Scanning electron micrographs of mature normal (A,D), _o2_ (B), and mod. _o2_ (C,E) R802 maize endosperm. A, B, and C represent soft (floury) regions; D and E represent hard (vitreous) regions. Bars represent 10 microns.

of protein bodies and other cytoplasmic features did not differ among the three genotypes (Fig. 10A, 10C, and 10E), protein bodies were easily distinguished by their size and staining characteristics. Both o2 and mod. o2 protein bodies (Fig. 10D and 10F) were smaller than those of normal maize (Fig. 10B). Whereas normal and o2 protein bodies were relatively homogeneous in staining density (Fig. 10B and 10D), mod. o2 protein bodies had multiple, non-concentric, circular, lightly-staining areas in an amorphous, darkly staining background (Fig. 10F). These R802 mod. o2 protein bodies appeared similar to mod. o2 protein bodies shown by Wallace et al. (1990b), who compared the morphology of W64A normal, W64A o2, and "Pool 25 QPM" mod. o2 from a different background. Whereas the heterogeneity of protein body staining is consistent with the markedly altered prolamin composition in mod. o2, the identity of the differentially staining regions could not be ascertained. However, Lending et al. (1988) reported immunocytochemical evidence that the darkly-staining region contains γ-zein, which is more abundant in mod. o2. This correlates with the higher relative amount of darkly-staining regions in mod. o2 compared to normal protein bodies (Fig. 10).

Examination of mature soft and hard regions of normal, o2, and mod. o2 R802 endosperms by scanning electron microscopy (Fig. 11), while confirming the microscopic appearance of hard and soft endosperm, did not provide additional insight into the mechanism of endosperm vitrification. Soft endosperm regions of the three genotypes were similar (Fig. 11A, 11B, and 11C), except for visibility of large protein bodies (appearing as light specks on the starch grains) in normal endosperm (Fig. 11A). The protein bodies of o2 and mod. o2 were presumably embedded in the dried cytoplasm adhering to the starch grains, but were not visible (Fig. 11B and 11C). Hard (vitreous) regions of normal and mod. o2 endosperm tissue were indistinguishable with this method (Fig. 11D and 11E).

CONCLUSIONS

Since mod. _o2_ kernels are phenotypically similar to normal kernels, we developed an RP-HPLC method of prolamin analysis that can identify and differentiate mod. _o2_ (QPM) maize. Major differences in accumulation of alcohol-soluble proteins in developing normal, _o2_ and mod. _o2_ occur. These differences in protein composition were supported by staining differences in transmission electron micrographs. Protein compositions and deposition may correlate with endosperm texture differences as well as with nutritional value, and be useful in identifying and breeding hard high-lysine maize.

LITERATURE CITED

Biemond, M. E. F., Sipman, W. A., and Olivie, J. 1979. Quantitative determination of polypeptides by gradient elution high pressure liquid chromatography. J. Liq. Chromatogr. 9:1407.

Das, O. P., and Messing, J. W. 1987. Allelic variation and differential expression at the 27-kilodalton zein locus in maize. Mol. Cell Biol. 7:4490.

Gentinetta, E., Maggiore, T., Salamini, F., Lorenzoni, C., Pioli, F., and Soave, C. 1975. Protein studies in 46 _opaque-2_ strains with modified endosperm texture. Maydica 20:145.

Landry, J., and Moureaux, T. 1970. Hétérogénéité des glutélines du grain de mais: extraction sélective et composition en acides aminés des trois fractions isolées. Bull. Soc. Chem. Biol. 52:1021.

Landry, J., Paulis, J. W., and Fey, D. A. 1983. Relationship between alcohol-soluble proteins extracted from maize endosperm by different methods. J. Agric. Food Chem. 31:1317.

Lending, C. R., Kriz, A. L., Larkins, B. A., and Bracker, C. E. 1988. Structure of maize protein bodies and immunocytochemical localization of zeins. Protoplasma 143:51.

Ludevid, M. O., Torrent, M., Martinez-Izquierdo, J. A., Puigdomenech, P., and Palau, J. 1984. Subcellular localization of glutelin-2 in maize (*Zea mays* L.) endosperm. Plant Mol. Biol. 3:227.

Misra, P. S., Mertz, E. T., and Glover, D. V. 1975. Studies on corn proteins. VI. Endosperm protein changes in single and double endosperm mutants of maize. Cereal Chem. 52:161.

Ortega, E. I., and Bates, L. S. 1983. Biochemical and agronomic studies of two modified hard-endosperm *opaque-2* maize (*Zea mays* L.) populations. Cereal Chem. 60:107.

Paulis, J. W., and Bietz, J. A. 1986. Separation of alcohol-soluble maize proteins by reversed-phase high performance liquid chromatography. J. Cereal Sci. 4:205.

Paulis, J. W., Bietz, J. A., Bogyo, T. P., Darrah, L. L., and Zuber, M. S. 1990. Expression of alcohol-soluble endosperm proteins in maize single- and double-mutants. Theor. Appl. Genet. 79:314.

Paulis, J. W., Bietz, J. A., Lambert, R. J., and Villegas, E. 1991. Identification of modified high-lysine genotypes by reversed-phase high performance liquid chromatography. Cereal Chem. 68:361.

Paulis, J. W., and Wall, J. S. 1971. Fractionation and properties of alkylated-reduced corn glutelin proteins. Biochim. Biophys. Acta 251:57.

Paulis, J. W., and Wall, J. S. 1977. Fractionation and characterization of alcohol-soluble reduced corn endosperm glutelin proteins. Cereal Chem. 54:1223.

Pedersen, K., Argos, P., Naravano, S. V. L. and Larkins, B. A. 1986. Sequence analysis and characterization of a maize gene encoding a high-sulfur zein protein of Mr 15,000. J. Biol. Chem. 261:6279.

Robutti, J. L., Hoseney, R. C., and Deyoe, C. W. 1974. Modified *opaque-2* corn endosperms. I. Protein distribution and amino acid composition. Cereal Chem. 51:163.

Ruskin, R. (ed). 1988. Quality-Protein Maize. National Academic Press, Washington, DC. p. 46.

Scopes, R. K. 1974. Measurement of protein by spectrophotometry at 205 nm. Anal. Biochem. 59:277.

Soave, C., Tardani, L., Di Fonzo, N., and Salamini, F. 1981. Zein level in maize endosperm depends on a protein under control of the opaque-2 and opaque-6 loci. Cell 27:403.

Sodek, L., and Wilson, C. M. 1971. Amino acid composition of proteins isolated from normal, opaque-2 and floury-2 corn endosperms by a modified Osborne procedure. J. Agric. Food Chem. 19:1144.

Wallace, J. C., Lopes, M. A., Paiva, E., and Larkins, B. A. 1990a. New methods for extraction and quantitation of zeins reveal a high content of γ-zein in modified opaque-2 maize. Plant Physiol. 92:191.

Wallace, J. C., Ohtani, T., Lending, C. R., Lopes, M., Williamson, J. D., Shaw, K. L., Gelvin, S. B., and Larkins, B. A. 1990b. Factors affecting physical and structural properties of maize protein bodies. In: Plant Gene Transfer, Lamb, C. J. and Beachy, R. N., eds., Alan R. Liss, Inc., New York, pp.205-216.

Wilson, C. M., Shewry, P. R., and Miflin, B. J. 1981. Maize endosperm proteins compared by sodium dodecyl sulfate gel electrophoresis and isoelectric focusing. Cereal Chem. 58:275.

TECHNIQUES FOR CHARACTERIZATION OF MAIZE
ENDOSPERM STORAGE PROTEINS AND IDENTIFICATION
OF NUTRITIONALLY IMPROVED MAIZE[1]

Curtis M. Wilson

National Center for Agricultural Utilization Research
Agricultural Research Service
U.S. Department of Agriculture
Peoria, IL 61604

INTRODUCTION

Maize is the highest yielding cereal and is noted
for starch production. However, maize supplies an
appreciable proportion of protein in the diet in many
societies around the world. The major storage
proteins, the Osborne zeins, lack lysine and
tryptophan and are low in cysteine and methionine,
four of the essential amino acids (Table I). Other
known maize seed proteins collectively provide
adequate proportions of essential amino acids. Thus
a more nutritious maize seed results when zein
synthesis is inhibited by the *opaque-2 (o2)*
mutation. Unfortunately, total protein, already low,
is often reduced in *o2* lines, while yield and
other agronomic properties are unfavorable.

---

[1]Presented at the Quality Protein Maize (QPM)
Symposium at the AACC 75th Annual Meeting in Dallas,
TX, Oct. 14-18, 1990.

Table I

Essential Amino Acid Patterns, mol % (Wilson, 1983)

Amino Acid	1973 FAO Provisional Pattern	Maize Non-zein Protein	Osborne Zein
Lysine	4.3	4.7	0.1
Threonine	3.9	4.9	3.0
Valine	4.2	7.1	3.6
Cysteine + Methionine	3.0	4.5	1.9
Isoleucine	3.5	4.0	3.8
Leucine	6.2	9.0	18.7
Phenylalanine + Tyrosine	4.0	6.2	8.7
Tryptophan	0.5	0.6	0
Total	29.6	41.0	39.8

Thus there have been attempts to find lines which maintain the desirable amino acid composition of maize *o2* lines while improving agronomic properties and retaining the protein content of non-mutant lines.

It is ironic that much research has been done on zein, the unwanted storage protein. We know nucleotide sequences, and hence amino acid sequences, for a number of zeins, and much is known about regulation of zein synthesis and of its deposition into protein bodies. The goal of maize improvement research is not merely to eliminate zein, but to replace it as the major storage protein by other proteins with desired amino acid compositions. Hopefully, knowledge about the deposition of zein as a storage protein may give us the means to increase the deposition of other more nutritious proteins in the corn kernel.

<u>Zein nomenclature</u>. I use an expanded definition of zein: "Zeins are alcohol-soluble proteins which occur principally in protein bodies of maize endosperm and which may or may not require reduction

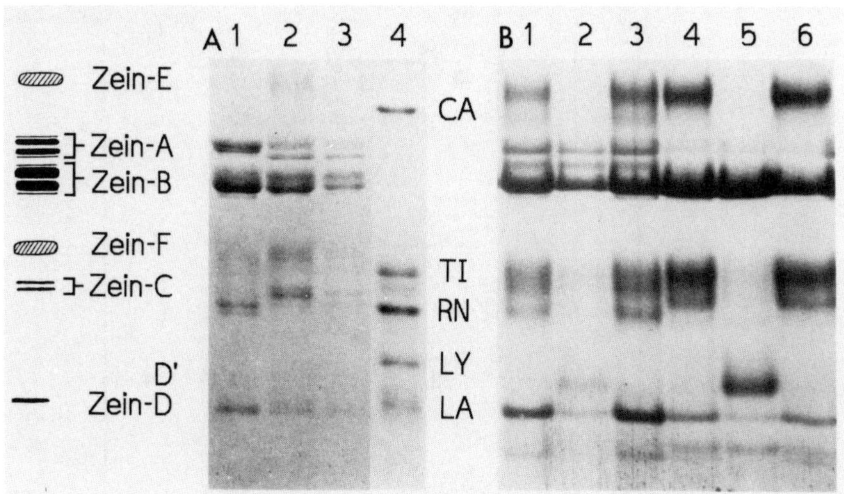

Fig. 1. Zein nomenclature, based on SDS-PAGE mobility (Wilson, 1986 in part). A. SDS-PAGE for three inbreds, showing two bands for C zein. Molecular mass standards: CA, carbonic anhydrase, 29 kD; TI, soybean trypsin inhibitor, 20.1 kD; RN, pancreatic ribonuclease, 13.7 kD; LY, lysozyme, 14.3 kD; and LA, α-lactalbumin, 14.2 kD. B. SDS-PAGE of W64A-*o2* (lanes 1-3) and Oh43-*o2* (lanes 4-6). Total zein extract, lanes 1 and 4; zein-1 (55% isopropanol extraction), lanes 2 and 5; and zein-2 (zein-1 precipitate extracted with 55% isopropanol and 2% mercaptoethanol), (lanes 3 and 6).

before extraction. Major classes of zeins are distinguished primarily by relative mobility (Rm) on SDS-PAGE and secondarily by solubility, amino acid composition (little or no lysine and tryptophan, relatively high contents of glutamine and proline), and other properties." This definition is based on characteristics of all alcohol-soluble proteins (Wilson, 1987). The nomenclature for these zeins is based on their relative mobilities on SDS-PAGE (Fig. 1). Zeins A and B are the traditional Osborne zeins, while zeins C-F are minor zeins which are extractable and soluble in alcoholic solutions only if reducing agents are present. In Table II I have

Table II

Zein Names, Derived from Their Different Properties (see text).

Zein Classes						
<u>Molecular Mass</u>						
Classes	A	B	C	D	E	F
From cDNA, kD	27	24	17	14	22	18
Commonly used	22	19	14, 15	10	27,28	16
Range, SDS-PAGE	21-26	18-24	13-17	9-14	27-31	15-21
<u>Solubility</u>						
Landry-Moureaux (1970)	Zein	Zein	G1	G1?	G2?	G1?
Paulis-Wall (1977)	Zein	Zein	WI-ASG	WI-ASG?	WS-ASG	WI-ASG
Esen (1987)	Alpha	Alpha	Beta	Delta?	Gamma	Beta
<u>Genetics</u>						
Thompson-Larkins (1989)	Alpha	Alpha	Beta	Delta	Gamma	Gamma
Chromosome[a]	4	4,7	6?	9	7	–
<u>HPLC Peak Area</u>						
Paulis-Bietz (1986)	4	4	1	4	2	3

[a] Benner et al, 1989, Wilson et al 1989.

144

reported molecular masses determined by SDS-PAGE by
various workers and the true molecular masses
obtained from the sequences determined by cDNA
cloning. I have proposed the letter nomenclature as
a filing system in which the known properties of each
individual zein can be collected (Wilson, 1987,
1991). The letters avoid the variable and confusing
numbers for apparent molecular masses applied to
zeins by different laboratories. Fig. 1A shows
inbreds in lanes 1 and 2 with sufficient protein to
show all of the zeins (including two C-zeins) and
lane 3 with less protein to show more details of the
A and B classes. Fig. 1B shows total zein (lanes 1
and 3), zein-1 (lanes 2 and 5), and zein-2 (lanes 3
and 6) fractions obtained from two *o2* lines. Low
levels of A zein are noted in relation to the other
zeins, as well as the lack of C-F zeins in the zein-1
extracts. Lane 5 has an as yet unknown protein (here
called D') migrating almost as fast as D zein. It is
probably not a D zein, for it stained violet with
Coomassie Blue R while D zein stains blue.

Other nomenclature. Different names for the zein
classes are collected in Table II. I don't use the
term glutelin, which signifies an extraction fraction
which is a mixture of unrelated proteins plus
previously unextracted zein (Sodek and Wilson,
1971). Even Osborne noted that only wheat glutenin
was accessible to satisfactory investigation
(Osborne, 1924). Although the Landry-Moureaux (1970)
and Paulis-Wall (1977) extraction systems greatly
improved our understanding of maize storage proteins,
clearcut differentiation of individual proteins by
extraction and solubility properties is difficult
because slightly changed conditions may put all or
part of a protein into a different extraction
fraction. Esen's nomenclature (1987) depends on
SDS-PAGE to identify the proteins separated by his
thorough manipulation of solubility properties. The
genetic classification (Thompson and Larkins, 1989)
differs slightly from the Esen solubility
classification, though both use a Greek letter
system. I find it useful to distinguish the two
major size classes of alpha zein as A and B, even

Table III
Amino Acid Contents (mol %)
Characteristic of Zein Classes

| | Zein Class | | | | | |
Amino Acid	A	B	C	D	E	F
Lysine[a]	0	0	0	0	0	0
Tryptophan[a]	0	0	0	0	0	0.6
Histidine	1-2	1-2	0	2	8[a]	2
Glutamine	21[a]	20[a]	16	12	16	19[a]
Proline	9	10	9	16	25[a]	15
Alanine	14[a]	14[a]	14[a]	5	5	7
Cysteine	0.5	1	4[a]	4[a]	7[a]	7[a]
Methionine	2	0.5	11[a]	22[a]	0-1	2
Leucine	19[a]	20[a]	10	12	9	8
Tyrosine	3	4	9[a]	1[a]	2[a]	6

[a] Amino acid levels which characterize one or more zeins. Data derived from Wilson (1987).

though they are genetically related. E and F zeins are also related genetically and share amino acid sequences, but they differ sufficiently in amino acid composition (Table III), size, and other properties that it is useful to classify them separately.

Amino acid content. Partial amino acid compositions, derived from cDNA sequences, show that each class of zein has certain amino acids which are especially high or low (Table III). These characteristic amino acids often signal the presence of a specific zein in extracts. The AB zeins have high glutamine, alanine and leucine. The minor zeins, C-F, all have moderate to high cysteine contents, resulting in formation of disulfide bonds and the requirement for reducing agents during extraction. C zein has high alanine, methionine, and tyrosine. D zein is noted for a very high methionine content, and also has low tyrosine. E zein has very high proline, high histidine, and low tyrosine. F zein is rich in glutamine, and has one tryptophan residue per molecule.

<u>Zein genes</u>.   The genes for AB zeins are found on chromosomes 4 and 7 (Wilson et al, 1989).   The structural gene for D zein is on chromosome 9, but the gene which regulates overproduction of this high methionine zein is on chromosome 4 (Benner et al, 1989).   The minor zeins produce three early peaks and the major zeins produce a late-eluting group by HPLC (Paulis and Bietz, 1986, Wilson, 1991). These will be discussed in detail below and by Dombrink-Kurtzman and Wilson (1991) (see also Paulis and Bietz 1986, Paulis et al. 1991). HPLC, a satisfactory system for the differentiation of normal inbreds (Wilson 1991), will also be shown to be satisfactory for identification of high-lysine hard endosperm lines.

MATERIALS AND METHODS

Maize endosperm was removed from seeds with a small drill.   Twenty milligrams was adequate. The samples were extracted with 5 μl per mg of a solution of 55% isopropanol, 5% mercaptoethanol, and 0.5% sodium acetate.   Sample size was about 20 μg of zein for IEF and analytical HPLC, 1 to 1.5 mg for preparative HPLC, and 6-20 μg for SDS-PAGE. The HPLC procedure is based on that of Paulis and Bietz (1986).   Reversed-phase HPLC separations were made with a segmented acetonitrile gradient (between 38.3 and 56% acetonitrile with 0.1% trifluoroacetic acid) to minimize analysis time while giving good peak separation.   Serial analyses consisted of taking individual zein bands or peaks isolated by one technique and assaying them by a second. Thus each zein could be recognized by position on IEF gels, relative mobility upon SDS-PAGE, and elution time by HPLC.   Full details for IEF, SDS-PAGE, and HPLC are given elsewhere (Wilson 1985, 1986, 1991).

RESULTS AND DISCUSSION

<u>Isoelectric focusing (IEF)</u>.   I previously reported results of a serial analysis procedure whereby individual zein bands separated by IEF were transferred to SDS-PAGE for an additional separation

(Wilson, 1985, 1986). This allowed about 70 individual zeins to be identified, and was useful in genetic studies of zein inheritance (Wilson et al, 1989). IEF separates the AB zeins quite well, but solubilities of the C, E, and F zeins cause difficulties. These electrophoretic methods depend upon protein stains for detection. Incomplete staining and variable color yields of zeins make quantitation difficult (Wilson 1984).

    HPLC patterns. The complete HPLC patterns at 210 nm and at 280 nm of all zeins in the inbred N28 are shown in Fig. 2. The absorbance at 210 nm is mostly due to peptide bonds, so that all proteins absorb

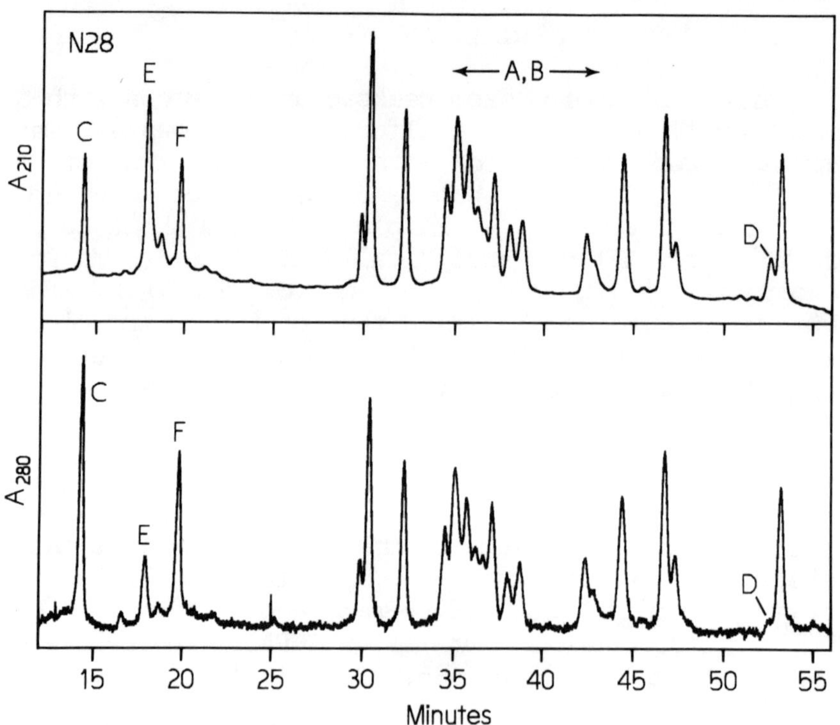

Fig. 2. HPLC of zeins of inbred N28, with a segmented gradient starting at 38.3% ACN. A_{210} shows D and E zeins, while A_{280} shows low absorption by D and E zeins (Wilson, 1991).

about equally (Buck et al, 1989). Absorbance at 280 nm depends upon tryptophan and tyrosine contents, which differ among the zeins (Table III). The relative heights of the AB zeins appear unchanged at 280 nm as compared to 210 nm. The C zein peak is greatly increased relative to the AB zeins at 280 nm, however. The F zein peak is also increased, the E zein peak is much reduced, and D zein is almost undetectable at 280 nm. Thus a dual wavelength detector allows minor zeins to be identified directly by the relative peak heights. HPLC appears to be suitable for quantitative analysis of all of the zeins.

    <u>HPLC-IEF analysis</u>. The AB zeins of inbred B57 separated into a large number of peaks. Serial analysis by preparative HPLC (not shown) followed by IEF of a few of the individual peaks (Fig. 3) revealed that zeins from peaks 6 and 11 moved to the same position by IEF, while a very large peak (divided into leading half A and trailing half B)

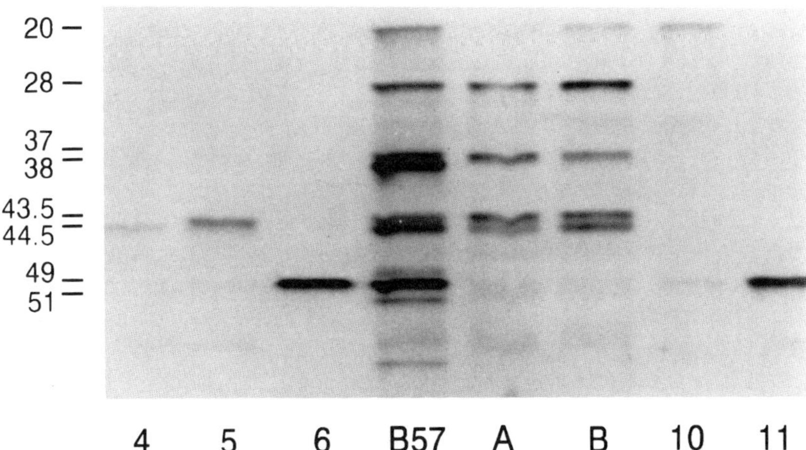

Fig. 3. Serial analysis of zeins isolated from inbred B57 by preparative HPLC, then run on IEF. Several individual peaks were assayed plus a large peak divided into parts A and B. The vertical scale is mm from the cathode of zein bands on a "typical" gel (Wilson, 1985).

consisted of zeins which separated into five bands by IEF. Supposedly identical IEF bands from different inbreds elute at different times when run on analytical HPLC (not shown). SDS-PAGE reveals a few more differences. Forty AB zeins were revealed by serial analysis of just four inbreds (Wilson 1991). Detailed analysis of the AB zeins is useful for genetic studies, for variety identification, and for detection of outcrossing during production of inbred and hybrid seed, but the major differences in mutant endosperms are seen in the other zein classes.

<u>Minor zeins</u>. The elution positions of the minor zeins were identified by collecting the first three peaks from a preparative HPLC column plus the penultimate peak, which had low absorbance at 280 nm (C, E, F, and D, respectively). The identities were confirmed by serial analysis on SDS-PAGE (Fig. 4). Two common genetic variants of C zein (C1 and C2) can be detected by SDS-PAGE (Fig. 1A, lanes 1 and 2) and HPLC (Wilson, 1991). The QPM samples had a small amount of a zein which appeared to be C zein by relative absorbance at 210 nm and 280 nm, but which eluted at different positions from the C zeins of corn belt lines. The identity of these proteins has not been confirmed. No confirmed variants of D, E or F zein have been detected by HPLC.

<u>Zein variations in maize germplasm</u>. A survey of zeins in several cornbelt inbreds and *o2* mutants is given in Table IV. As expected, the AB (Osborne) zeins predominate in normal inbreds. There is a wide variation in the contents of the high methionine C and D zeins. It might be possible to utilize this variation for production of lines with improved nutritional value. The E and F zeins occur in larger amounts, with up to three-fold variations in percentage. The *o2* mutation reduced the AB zein content of R802 and increased all of the minor zeins, especially F zein. As these zeins are concentrated in the periphery of the protein bodies (Wallace et al, 1990), these changes may be related to some aspect of protein body development in *o2* endo- sperms. The modified R802-*o2* endosperm (obtained from R. Lambert, Univ. of Illinois) was marked by a

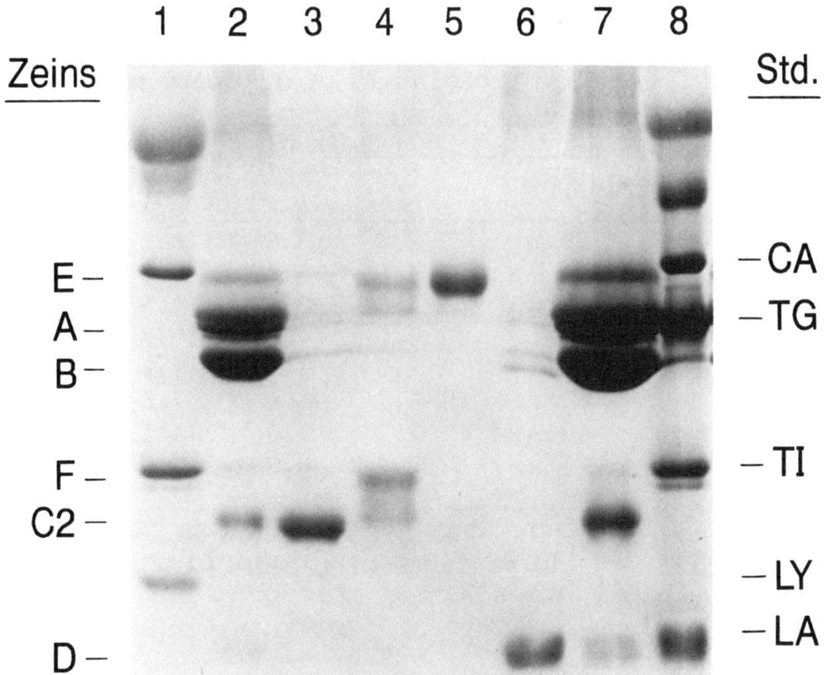

Fig. 4.   SDS-PAGE of C, D, E, and F zein fractions
from preparative HPLC.   Lane 1,   Bio Rad standard
proteins; lane 2, A619, whole zein; lane 3, C2 zein;
lane 4, F zein; lane 5, E zein; lane 6, D zein; lane
7, A619 whole zein; lane 8, Sigma protein standards.
See Fig. 1 for identification of standard proteins,
plus TG, trypsinogen, 24 kD (Wilson, 1991).

very high level of E zein, over four times the level
of the unmodified *o2* R802.   The C zein in this
modified version was the same type as in the normal
inbred, while the C and D levels were similar to
normal R802.   Several CIMMYT QPM populations were
sampled, with the average values for the zeins
resembling that of the R802-mod. *o2*.   The very
high levels of zein E appear to be characteristic of
modified hard endosperm *o2* lines derived from two
independent sources.   Whether or not the variations
in E zein contents of the corn belt inbreds influence

Table IV
Zein Classes as Percentages of Total
Alcohol-soluble Proteins (a preliminary survey).

Line	AB	C		D	E	F
W64A	85.5	C1	2.7	1.0	6.5	4.3
Oh43	78.9	C2	4.0	3.4	8.6	5.1
Mo17	78.6	C1	1.5	0.9	14.9	4.1
BSSS53	72.1	C1	4.8	4.8	10.7	7.6
R802-N	85.8	C1	2.5	1.7	4.9	3.2
R802-o2	60.1	C1	4.1	5.6	7.5	12.4
R802-mod.o2	49.9	C1	2.2	1.3	34.1	12.6
QPM[a]	46.1	C?	2.4	0.9	40.0	11.0

[a] Average of two samples each of four populations of CIMMYT QPM lines, with a variable unidentified C-zein. Details elsewhere in this book.

kernel hardness is unknown. Detailed studies on the variations in the zeins of QPM and other lines are reported elsewhere in this book.

Conclusions. The serial analysis procedure using IEF, SDS-PAGE, and HPLC allows many individual zeins to be identified. HPLC alone, after a simple extraction with alcohol, a reducing agent, and sodium acetate, gives quantitative analyses of all of the zeins. QPM lines have exceptionally high levels of E zein.

LITERATURE CITED

Benner, M. S., Phillips, R. L., Kirihara, J. A., and Messing, J. W. 1989. Genetic analysis of methionine-rich storage protein accumulation in maize. Theor. Appl. Genet. 78:761-767.

Buck, M. A., Olah T. A., Weitzmann C. J., and Cooperman, B. S. 1989. Protein estimation by the product of integrated peak area and flow rate. Anal. Biochem. 182:295-299.

Dombrink-Kurtzman, M. S., and Wilson, C. M. 1991. Characterization of proteins in hard and soft endosperm of quality protein maize. These proceedings.

Esen, A. 1987. A proposed nomenclature for alcohol-soluble proteins (zeins) of maize (*Zea mays* L.). J. Cereal Sci. 5:117-128.

Landry, J., and Moureaux, T. 1970. Heterogeneite des glutelines du grain de mais: extraction selective et composition en acides amines des trois fractions isolees. Bull Soc. Chim. Biol. 52:1021-1037.

Osborne, T. B. 1924. The Vegetable Proteins, 2nd ed. Longmans, Green and Co., London. 154 pp.

Paulis, J. W., and Bietz J. A. 1986. Separation of alcohol-soluble maize proteins by reversed-phase high performance liquid chromatography. J. Cereal Sci. 4:205-216.

Paulis, J. W., Bietz, J. A., Lambert, R. J., and Villegas, E. M. 1991. Identification of modified high-lysine maize genotypes by reversed-phase high-performance liquid chromatography. Cereal Chem. 68:361-365.

Paulis, J. W., and Wall, J. S. 1977. Fractionation and characterization of alcohol-soluble reduced corn endosperm glutelin proteins. Cereal Chem. 54: 1223-1228.

Sodek, L., and Wilson, C. M. 1971. Amino acid composition of proteins isolated from normal, *opaque-2*, and *floury-2* corn endosperms by a modified Osborne procedure. J. Agric. Food Chem. 19:1144-1150.

Thompson, G. A., and Larkins, B. A. 1989. Structural elements regulating zein gene expression. BioEssays 10:108-113.

Wallace, J. C., Lopes, M. A., Paiva. E., Larkins, B. A. 1990. New methods for extraction and quantitation of zeins reveal a high content of γ-zein in modified *opaque-2* maize. Plant Physiol 92:191-196.

Wilson, C. M. 1983. Seed protein fractions of maize, sorghum, and related cereals. Pages 271-307 in: "Seed Proteins: Biochemistry, Genetics, Nutritive Value", W. Gottschalk and H.P. Muller, eds. M. Nijhoff/Junk, The Hague.

Wilson, C. M. 1984. Staining of proteins on gels: Comparisons of dyes and procedures. Pages 236-247 in: "Methods in Enzymology, Enzyme Structure, Part I., Vol. 91." C. H. W. Hirs and S. N. Timasheff, eds. Academic Press, New York.

Wilson, C. M. 1985. A nomenclature for zein polypeptides based on isoelectric focusing and sodium dodecyl sulfate polyacrylamide gel electrophoresis. Cereal Chem. 62:361-365.

Wilson, C. M. 1986. Serial analysis of zein by isoelectric focusing and sodium dodecyl sulfate gel electrophoresis. Plant Physiol. 82:196-202.

Wilson, C. M. 1987. Proteins of the kernel. Pages 273-310 in: "Corn: Chemistry and Technology." S. A. Watson and P. E. Ramstad, eds. Am. Assoc. Cereal Chem., St. Paul, MN.

Wilson, C. M. 1991. Multiple zeins from maize endosperms characterized by reversed-phase high performance liquid chromatography. Plant Physiol. 95:777-786.

Wilson, C. M., Sprague, G. F., and Nelsen, T. C. 1989. Linkages among zein genes determined by isoelectric focusing. Theor. Appl. Genet. 77:217-226.

CHARACTERIZATION OF PROTEINS IN HARD AND SOFT
ENDOSPERM OF QUALITY PROTEIN MAIZE[1]

Mary Ann Dombrink-Kurtzman and Curtis M. Wilson

National Center for Agricultural Utilization Research
Agricultural Research Service,
U. S. Department of Agriculture[2],
Peoria, IL 61604

INTRODUCTION

In 1964 Mertz et al. reported that _opaque-2_ (_o2_)
maize mutant kernels possessed twice the normal level
of lysine (Mertz et al., 1964). The improved protein
quality, however, was accompanied by undesirable
characteristics, including decreased yield, soft
texture endosperm, increased kernel moisture at
maturity and greater susceptibility to insect and
mold damage (Mertz, 1986). Through introduction of
genetic modifiers, _o2_ maize with hard kernels and
better nutritional quality than normal maize has been
developed (Ortega and Bates, 1983; Quality Protein
Maize, 1988).

---

[1]Presented at the 75th Annual Meeting of the
American Association of Cereal Chemists, Dallas, TX,
October 14-18, 1990.
[2]The mention of firm names or trade products does
not imply that they are endorsed or recommended over
other firms or similar products not mentioned.

The quality protein maize (QPM) program at CIMMYT (International Maize and Wheat Improvement Center) has been involved in development and improvement of gene pools and populations with high yield potential and kernels having hard endosperm and increased amounts of the essential amino acids lysine and tryptophan (Quality Protein Maize, 1988). Research has focused on maintaining the recessive mutant (o2) background.

Two Types of Endosperm

A representation of a corn kernel (Hoseney, 1986) is shown in Figure 1. The kernel has both horny and floury endosperm. The horny portion is hard, vitreous and translucent, while the floury part is soft and opaque. Many studies have described differences between these two types of endosperm (Christianson, 1970; Subramanyam et al., 1980; Wolf et al., 1952, 1967, 1969; Wolf and Khoo, 1970). The floury portion has larger cells due to cell enlargement in the inner area, concomitant with cell division in the outer area. Cell walls are thicker in floury endosperm. Breakage of kernels occurs

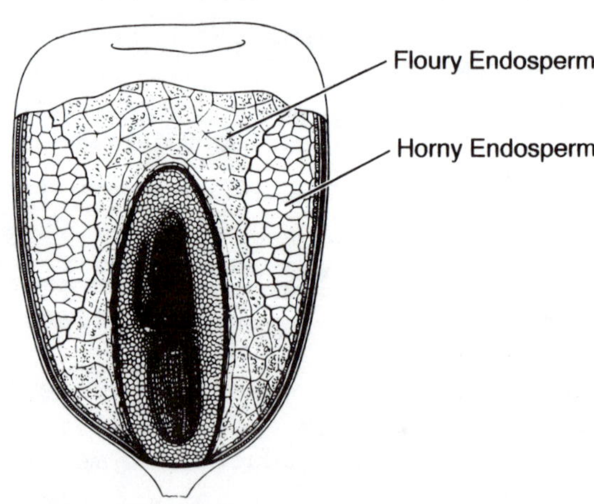

Fig. 1. Longitudinal section of a corn kernel. Courtesy of Corn Refiners Association, Inc., Washington, DC.

through cells in the floury portion, but along cell walls in the horny portion. Protein bodies, which contain the alcohol-soluble storage proteins, zeins, are both larger and more numerous in the horny endosperm. Spherical starch granules with space between them occur in floury endosperm, whereas horny endosperm contains compacted, polygonal starch granules, with abundant, clinging protein matrix (Robutti et al., 1974). The junction of hard and soft endosperm within a kernel is abrupt.

The many differences between horny and floury endosperm suggest a fundamental difference between the two types of cells. Such cells could represent different cell lineages or different stages of differentiation, with cells in the horny endosperm being more highly differentiated. Alternatively, "hardness" could be due, in the simplest case, to the presence of a specific "hardness protein", or the absence of a "softness protein."

We undertook the present study to determine the relationship between maize protein composition and endosperm hardness/softness. Our assumption has been that maize protein composition and distribution directly influence endosperm texture and physical properties, thereby affecting milling and susceptibility toward breakage.

MATERIALS AND METHODS

Individual kernels were hand dissected and portions differing in texture were examined. Pericarp and germ were removed after soaking kernels in distilled water for five minutes. Kernels were allowed to dry overnight. The soft portion was removed with a Dremel Moto-Tool (Emerson Electric Company) and the hard portion was ground in a WIG-L-BUG grinder (Crescent Dental Manufacturing Company). Approximate grinding time for the hard portion was three minutes per kernel. Both fractions were put through a 250 μm mesh screen. Fractions from ten kernels were pooled for analysis. Duplicate extractions were performed. The percent hard endosperm was determined by weight.

Samples Studied

Samples included QPM populations 61-70 (supplied by M. Bjarnason, CIMMYT); QPM F337 (supplied by L. Rooney, Texas A&M University); modified o2 lines Elite Single Cross Synthetic (El.Sc.Syn.), Synthetic Disease Oil (Syn.D.O.) and experimental hybrid 403 (supplied by R. J. Lambert, University of Illinois), plus Crow's high lysine corn (Crow's Hybrid Corn Company, Milford, IL) and F2, a normal French flint inbred (Pfister Hybrid Corn Company, El Paso, IL). Corn Belt hybrids and inbreds were also examined for comparison.

Extraction of Proteins

Hard and soft fractions of endosperm were extracted separately with SPE/DTT (0.5 M NaCl, 50 mM NaH_2PO4, 10 mM Na_2EDTA, pH 7.8; 0.2% dithiothreitol) or with alcohol (70% ethanol, 5% 2-mercaptoethanol, 0.5% Na acetate) to obtain salt- and alcohol-soluble proteins, respectively. Endosperm fractions were extracted in 1.5 ml polypropylene tubes at room temperature by shaking on a Buchler Vortex-Evaporator for two hours, using 20 mg sample per 200 μl SPE/DTT and 50 mg per 250 μl alcohol solution.

Analysis of Extracts

Following centrifugation (Eppendorf centrifuge), supernatants were analyzed by sodium dodecyl sulfate polyacrylamide gel electrophoresis (SDS-PAGE) and reversed-phase high-performance liquid chromatography (RP-HPLC). Extracts were mixed with an equal volume of sample buffer and analyzed by SDS-PAGE, using a modification of the Laemmli procedure (Fling and Gregerson, 1986).

For RP-HPLC, alcohol-soluble proteins were diluted 1:10 with 55% isopropanol, 5% 2-mercapto-ethanol; salt-soluble proteins were undiluted. Proteins were separated on a 250 mm x 4.6 mm Vydac RP-column (C_{18}), 5 μm, 300 Å, using a nonlinear 60 min gradient [25-64.4% acetonitrile (CH_3CN), in the presence of 0.1% trifluoroacetic acid], followed isocratically at 64.4% CH_3CN for 5 min. For optimal separation of all protein peaks, the following gradient was used: Starting with 25% CH_3CN, the concentration of CH_3CN was increased 1%/min for 25 min, followed by increases of 0.2%/min for 30 min and then 1.68%/min for 5 min. The column was operated at $55^{\circ}C$ with a flow rate of 1 ml/min. Twenty microliter samples were injected, using a Waters Associates WISP 710B automatic sample injector and a Spectra-Physics SP8700 Solvent Delivery System. Absorbance was monitored at 210 nm (absorbance range = 0.2). Data were stored and processed on a ModComp computer system.

RESULTS AND DISCUSSION

Alcohol-Soluble Maize Storage Proteins

Various nomenclatures are used for alcohol-soluble maize storage proteins. That used here is shown in a RP-HPLC chromatogram (Fig. 2) of the alcohol-soluble proteins from hard and soft endoperm of a non-QPM line, F2 (French flint inbred). These and all comparisons were of equal weights of endosperm. Peaks are labeled 1, 2, 3 and 4 (Paulis and Bietz, 1986), with the most hydrophobic proteins eluting last. Numbers in parentheses are commonly used apparent molecular masses, from SDS-PAGE (Larkins et al., 1989; Thompson and Larkins, 1989). Zein classes α, β and γ are based on differential solubility in aqueous alcohol solutions (Esen, 1987) or on genetic classes determined by sequence homology (Geraghty et al., 1982; Pedersen et al., 1986; Prat et al., 1987). Peak area 4 contained the α-zeins (19 and 22 kD), located in the core of protein bodies (Lending et al., 1988; Lending and Larkins, 1989), and also the 10 kD δ-zein (Kirihara et al., 1988). The opaque-2 mutation is known to reduce the amount of α-zeins by approximately 50% (Mertz et al., 1964). Peak 1 contained β-zein (15 kD), and gamma-zeins (27 and 16 kD) were in peaks 2 and 3, respectively. The β- and γ-zeins are found at the periphery of

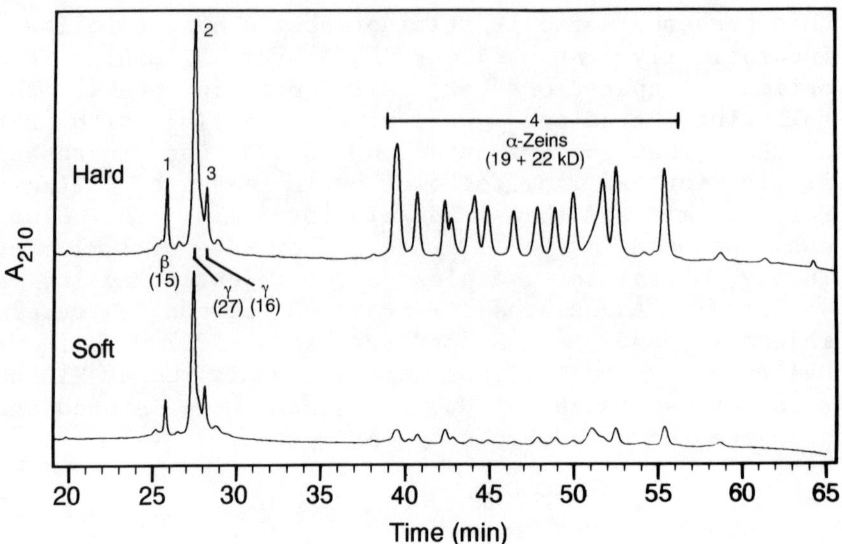

Fig. 2.   RP-HPLC of alcohol-soluble proteins from hard and soft endosperm from F2, a French flint inbred.   Total zeins were extracted by alcohol solutions containing a reducing agent and sodium acetate and analyzed using a C_{18} column and a nonlinear 25-64.4% acetonitrile gradient for 60 min. Numbers in parentheses are apparent molecular masses, based on SDS-PAGE.   The Greek letters α, β and γ refer to genetic classes, based on sequence homology.

protein bodies (Lending et al., 1988; Lending and Larkins, 1989).

Data for F2, a normal French flint inbred (Table I) was typical of nonmutant lines. Greater than 70% of alcohol-soluble proteins in hard endosperm were α-zeins.   The relative percentage of peak 2 (γ-zein) in soft endosperm was more than twice that in hard endosperm, and resembled the percentages in QPM fractions.   The results correlated with the model of protein body development proposed by Lending and Larkins (1989), who examined endosperm tissue using immunolocalization techniques with light and electron microscopy.

TABLE I

Relative Areas and Percentages of
Alcohol-Soluble Maize Endosperm Proteins

Sample	Texture	RP-HPLC Fractions			
		1	2	3	4
F337	Hard	2,000[a]	58,200	19,500	44,900
		1.4[b]	39.1	12.9	29.6
	Soft	1,200	43,300	12,500	23,800
		1.2	41.7	12.0	22.8
El.Sc. Syn.	Hard	2,800	57,200	25,300	67,200
		1.6	33.3	14.7	38.6
	Soft	2,100	49,800	15,700	36,600
		1.6	39.0	12.2	28.4
Syn.D.O.	Hard	2,200	84,100	25,000	82,600
		1.0	38.5	11.5	37.9
	Soft	2,200	64,200	16,900	33,800
		1.6	45.1	11.8	23.9
403	Hard	2,600	63,600	25,000	73,100
		1.4	34.4	13.5	39.4
	Soft	1,200	52,900	15,700	22,100
		1.1	49.0	14.5	20.4
Crow's High Lysine	Hard	2,400	27,200	12,600	42,700
		2.6	29.0	13.5	45.7
	Soft	600	27,500	10,800	26,000
		0.9	39.3	15.4	37.2
F2 French Flint	Hard	8,300	47,400	9,400	208,800
		2.9	16.2	3.2	71.4
	Soft	4,300	30,700	7,700	29,300
		5.0	35.7	8.9	33.6

[a]Relative area.
[b]Relative percentage.

Developmental Changes

Early in development, protein bodies had both β- and γ-zeins, but little or no α-zeins (similar to F2 soft endosperm). In the final stages of protein body maturation, α-zeins (present in high amounts seen in F2 hard endosperm) filled most of the core of the protein body and were surrounded by a thin layer of β- and γ-zeins. For F2 and flint QPM populations, 90 percent of the endosperm was hard. The amount of hard endosperm in dent QPM populations was 80 percent, whereas Corn Belt lines averaged 70 percent.

Zeins in Hard and Soft Endosperms

We observed two main differences between hard and soft endosperm from QPM populations (Table II). First, hard endosperm had more alcohol-soluble proteins and a greater amount of peak area 4 (α-zeins) proteins than did soft endosperm in most populations. Secondly, soft endosperm had a higher proportion of peak 2 (γ-zein), as well as a very low amount of peak area 4 proteins, than did hard endosperm (Table II).

In contrast to hard and soft endosperm of normal maize (Fig. 2), alcohol-soluble protein compositions of hard and soft endosperm of QPM population 63 (Fig. 3) were very similar, with low peak area 4 and high peak 2. Relative areas of all zeins were slightly higher in hard, than in soft endosperm (Table II). This may indicate that protein bodies in the hard and soft endosperm of population 63 are of similar composition, but the number of protein bodies in the hard endosperm is greater.

Relative areas and percentages of alcohol-soluble proteins from hard and soft endosperm of QPM populations 61-70 are shown in Table II. Percentages did not add up to 100 because unidentified peaks were not counted. For QPM populations 63 and 64, relative percentages of RP-HPLC fractions were similar in the hard and soft endosperm. In QPM population 62, however, peak area 4 was three-fold greater in the hard endosperm,

TABLE II

Relative Areas and Percentages of
Alcohol-Soluble Maize Endosperm Proteins

Sample	Texture	RP-HPLC Fractions		
		2	3	4
POP 61	Hard	37,900[a]	11,000	34,200
		38.8[b]	11.3	35.2
	Soft	28,000	6,100	21,500
		42.3	9.3	32.5
POP 62	Hard	34,200	12,700	33,700
		36.5	13.5	36.0
	Soft	25,800	8,100	11,000
		41.8	13.2	17.7
POP 63	Hard	32,100	9,700	25,400
		40.0	12.0	31.6
	Soft	28,400	7,900	22,500
		38.5	10.7	30.5
POP 64	Hard	34,700	10,700	28,100
		41.7	12.9	33.8
	Soft	25,500	7,000	20,800
		40.5	11.1	33.0
POP 65	Hard	45,600	9,000	30,600
		46.5	9.2	31.3
	Soft	31,400	7,200	3,600
		57.1	13.1	6.6
POP 66	Hard	38,300	10,900	24,900
		42.6	12.1	27.7
	Soft	27,200	5,800	19,500
		44.4	9.5	31.8
POP 67	Hard	38,300	9,900	29,900
		40.4	10.4	33.0
	Soft	25,500	5,500	2,300
		54.3	11.7	5.0
POP 68	Hard	28,800	7,400	25,000
		38.0	9.8	33.1
	Soft	22,700	4,700	10,000
		45.6	9.4	20.3
POP 69	Hard	40,800	11,300	31,400
		40.0	11.2	31.2
	Soft	20,200	4,400	5,400
		48.3	10.6	12.8
POP 70	Hard	31,900	9,500	20,600
		41.2	12.3	26.7
	Soft	23,800	5,800	14,500
		42.9	10.4	25.9

[a]Relative area.
[b]Relative percentage.

163

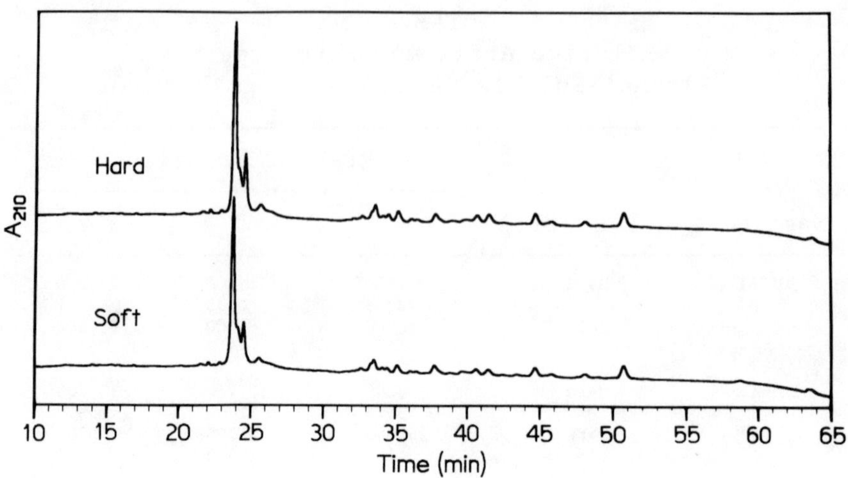

Fig. 3. RP-HPLC of alcohol-soluble proteins from hard and soft endosperm from QPM population 63. Extraction and chromatographic conditions were as described for Fig 2.

whereas relative percentage of peak 2 was greater in soft endosperm. Peak 1 (β-zein) was absent in all CIMMYT QPM populations.

Somewhat different profiles were observed for other QPM populations, possibly due to the presence of different modifier genes among the QPM populations. Soft endosperm of QPM population 65 (Fig. 4) had a very low amount of peak area 4 (α-zeins) and higher relative percentages of peaks 2 and 3 (γ-zeins). This type of profile was also seen for QPM populations 67 and 69. All three are flint type (CIMMYT International Maize Testing Program, 1985). QPM populations 65, 67 and 69 all had a much greater (6 to 10-fold) peak area 4 in hard endosperm (Table II), indicating that protein body composition may differ between hard and soft endosperm.

The hard endosperm from F337 (an open pollinated, experimental QPM grown in Texas) and modified o2 corn (from University of Illinois) (Table I) had twice as much peak area 4 protein as did soft

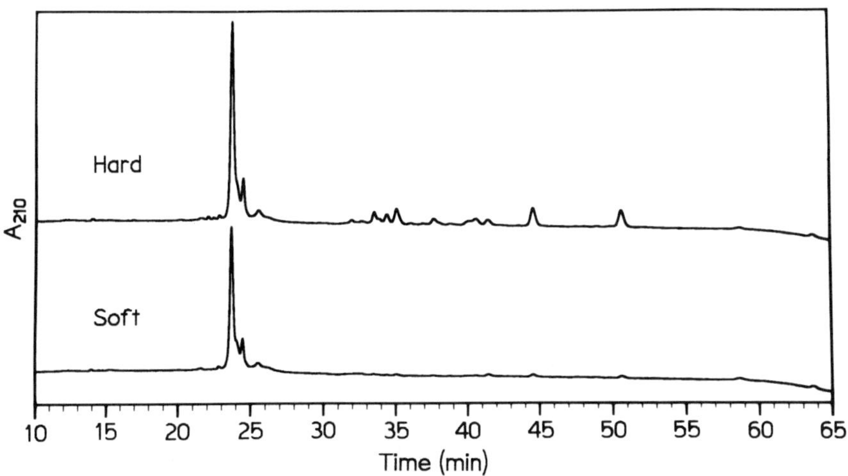

Fig. 4. RP-HPLC of alcohol-soluble proteins from hard and soft endosperm from QPM population 65. Extraction and chromatographic conditions were as described for Fig. 2.

endosperm. The relative percentage of peak 2 was higher in soft endosperm than in hard endosperm. Relative percentages of peak 2 for these samples were similar to those of QPM populations 61-70. The amount of the 27 kD γ-zein has been reported to be dramatically increased in QPM varieties (Wallace et al., 1990; Paulis et al., 1991).

SDS-PAGE analyses of alcohol-soluble proteins from F2, Crow's high lysine corn and QPM populations 69 and 70 are shown in Figure 5. This gel was purposely overloaded to determine if any proteins in addition to α-, β- and γ-zeins were present. Gels with lesser amounts of proteins were also run. Preparative RP-HPLC, followed by SDS-PAGE, aided in identification of proteins present in peaks 1-3. Faint bands at 43 and 55 kD were seen for non-QPM samples, but QPM populations had a prominent 55 kD band. This band is in the molecular size range of the 50 and 58 kD protein body-associated reduced soluble proteins reported by Vitale et al. (1982) to contain lysine.

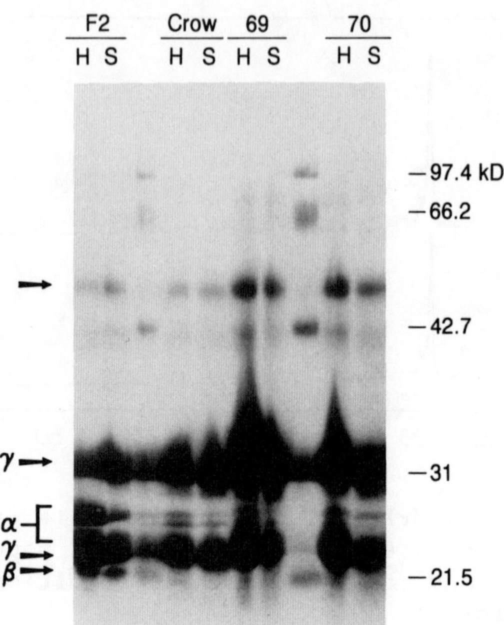

Fig. 5.   SDS-PAGE of alcohol-soluble proteins from hard (H) and soft (S) endosperm from F2, Crow's high lysine and QPM populations 69 and 70. The molecular mass standards (Bio-Rad) in lanes 3 and 8 were phosphorylase b (97.4 kD), BSA (66.2 kD), ovalbumin (42.7 kD), carbonic anhydrase (31 kD), soybean trypsin inhibitor (21.5 kD) and lysozyme (14.4 kD). Genetic classes are indicated on the left.

For QPM populations 69 and 70 (Fig. 5) [also populations 61-68 (data not shown)], the amount of α-zeins (mainly 22 kD) was low and β-zein (15kD) was absent. The predominant proteins were γ-zeins (16 and 27 kD). F2 had normal amounts of α- and β-zeins. In each genotype, soft endosperm fractions had less total zein than did hard endosperm.

Differences existed in proportions of peaks 2 and 4 (γ- and α-zeins, respectively) in hard and soft endosperm from many samples. Therefore, when comparing kernels, if only a small amount of endosperm is removed for analysis, care must be taken to note if the sample is from hard or soft endosperm.

Salt-Soluble Proteins

Figure 6 compares elution profiles of alcohol and salt extracts of QPM population 70 hard endosperm. Salt-soluble proteins eluted earlier than alcohol-soluble proteins. The main peak at 24 minutes in the salt-soluble fraction did not correspond to any peak in the alcohol-soluble fraction. In QPM population 67 (Fig. 7), this prominent peak was twice as abundant in hard as in soft endosperm. Salt-extracts of hard and soft endosperm from QPM populations POB 61-70 (Table III) and from non-QPM varieties (data not shown) all showed this same relationship.

When examined by SDS-PAGE (Fig. 8), all samples of salt-soluble proteins from soft endosperm from QPM populations 61-68 contained a 50 kD protein absent in hard endosperm. [Soft endosperm from QPM populations 69 and 70 (data not shown) also had a 50 kD protein.]

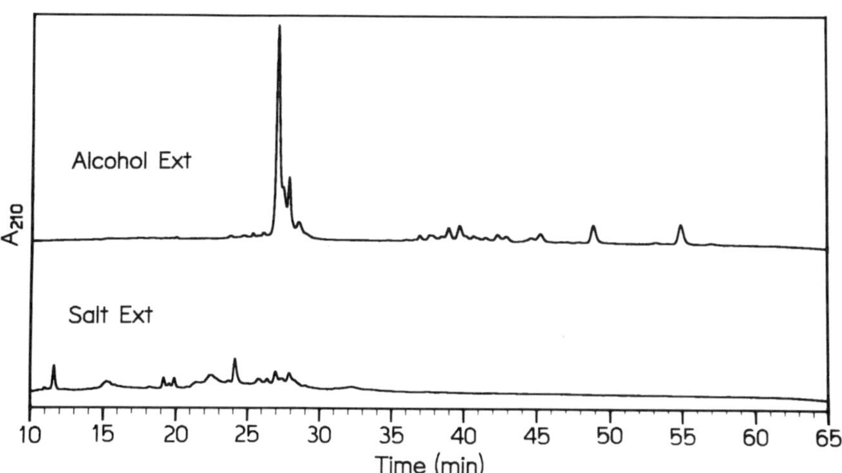

Fig. 6. RP-HPLC of alcohol- and salt-soluble proteins from QPM population 70 hard endosperm. Salt-soluble proteins were extracted in phosphate buffer containing NaCl, EDTA and DTT. Alcohol extraction and chromatographic conditions were as described for Fig 2.

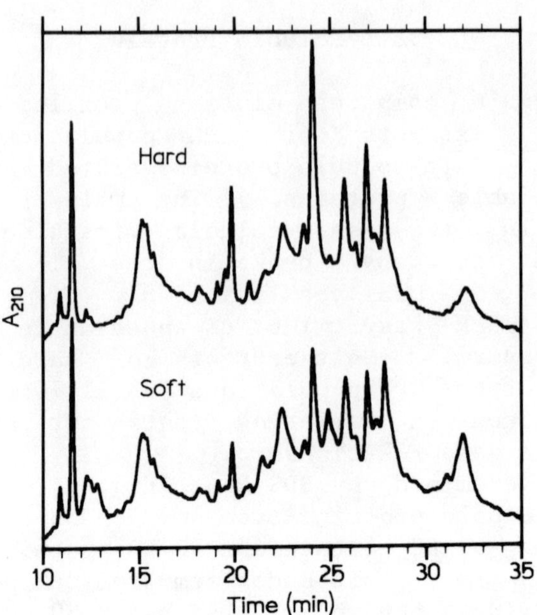

Fig. 7.   RP-HPLC of salt-soluble proteins from hard and soft endosperm from QPM population 67. This represents an enlargement compared to Fig. 6. Extraction and chromatographic conditions were as described for Fig. 6.

A faint 95 kD protein was more pronounced in hard endosperm.   Perhaps this latter band corresponded to the RP-HPLC peak present in two-fold greater amount in hard than in soft endosperm.

CONCLUSIONS

Hard and soft endosperm from normal and QPM kernels differed in composition of both salt- and alcohol-soluble protein fractions. QPM samples were composed mainly of γ-zeins (16 and 27 kD). Soft endosperms contained increased percentages of γ-zein (27 kD) compared to hard endosperms. Conversely, hard endosperms contained higher total amounts of alcohol-soluble proteins and up to ten

TABLE III

Relative Percentages of Prominant Peak[a] of
Salt-Soluble Maize Endosperm Proteins

| | Texture | |
Sample	Hard	Soft
POP 61[b]	8.7	5.7
POP 62	14.1	6.4
POP 63	9.0	6.3
POP 64	15.8	7.3
POP 65	15.7	7.5
POP 66	12.8	6.7
POP 67	15.2	8.7
POP 68	14.0	7.1
POP 69	17.0	7.3
POP 70	14.2	7.4

[a]Peak eluted at 24 minutes from RP-HPLC. Salt-soluble proteins were extracted in phosphate buffer containing NaCl, EDTA and DTT.
[b]Population is abbreviated as POP.

times the amounts of α-zeins present in soft endosperms. All QPM samples contained an alcohol-soluble 55 kD protein that was present in much reduced levels in normal (non-QPM) Corn Belt lines. Also, a salt-soluble 50 kD protein was observed only in the soft endosperm of all QPM populations. RP-HPLC of salt-soluble proteins revealed a prominent peak present in twice the amount in hard endosperm as in soft endosperm. These differences in protein composition may relate to composition of protein bodies and to texture of the endosperm from which samples were obtained.

169

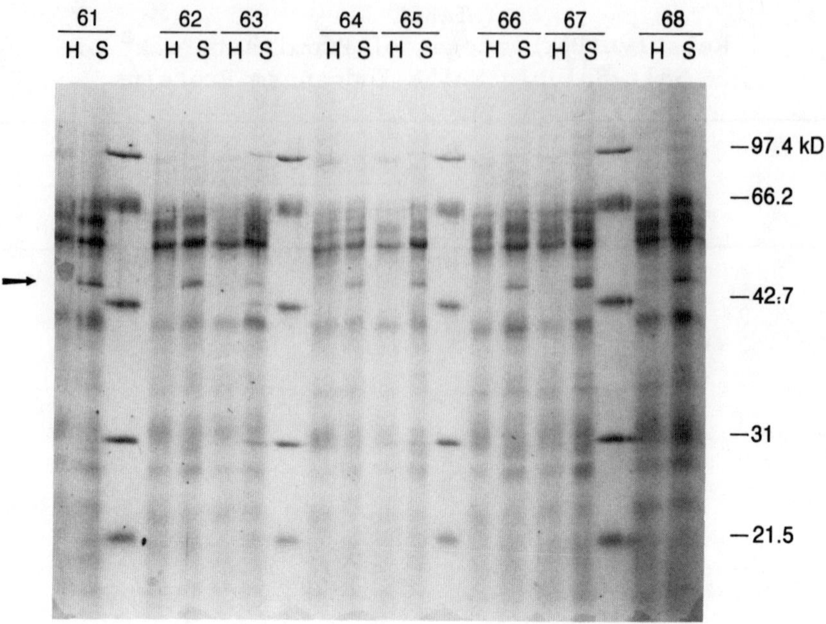

Fig. 8. SDS-PAGE of salt-soluble proteins from hard (H) and soft (S) endosperm from QPM populations 61-68. Arrow indicates band present only in soft endosperm. Molecular weight markers are in lanes 3, 8, 13 and 18 and were as described for Fig. 5.

LITERATURE CITED

Christianson, D. D. 1970. Genetic variation in cereal grains and processing effects on cereal flours as evaluated by scanning electron microscopy. Pages 161-168 in: Proc. 3rd Annual Scanning Electron Microscopy Symposium, Chicago, IL.

CIMMYT International Maize Testing Program. 1985 Final Report. 1985. CIMMYT, Mexico, D.F. Mexico.

Esen, A. 1987. A proposed nomenclature for the alcohol-soluble proteins (zeins) of maize (_Zea mays_ L.) J. Cereal Sci. 5:117.

Fling, S. P., and Gregerson, D. S. 1986. Peptide and protein molecular weight determination by electrophoresis using a high-molarity Tris buffer system without urea. Anal. Biochem. 155:83.

Geraghty, D. E., Messing, J., and Rubenstein I. 1982. Sequence analysis and comparison of cDNAs of the zein multigene family. EMBO J. 1:1329.

Hoseney, R. C. 1986. Principles of Cereal Science and Technology. American Association of Cereal Chemists, St.Paul, MN.

Kirihara, J. A., Petri, J. B., and Messing, J. 1988. Isolation and sequence of a gene encoding a methionine-rich 10-kDa zein protein from maize. Gene 71:359.

Larkins, B. A., Lending, C. R., Wallace, J. C., Galili, G., Kawata, E. E., Geetha, K. B., Kriz, A. L., Martin, D. N., and Bracker, C. E. 1989. Zein gene expression during maize endosperm development. Pages 109-120 in: The Molecular Basis of Plant Development. R. Goldberg, ed. Alan R. Liss, Inc., New York.

Lending, C. R., Kriz, A. L., Larkins, B. A., and Bracker, C. E. 1988. Structure of maize protein bodies and immunocytochemical localization of zeins. Protoplasma 143:51.

Lending, C. R., and Larkins, B. A. 1989. Changes in the zein composition of protein bodies during maize endosperm development. Plant Cell 1:1011.

Mertz, E. T. 1986. Genetic and biochemical control of grain protein synthesis in normal and high lysine cereals. Pages 222-262 in: World Review of Nutrition and Dietetics. G.H. Bourne, ed. S. Karger, Basel.

Mertz, E. T., Bates, L. S., and Nelson, O. E. 1964. Mutant gene that changes protein composition and increases lysine content of maize endosperm. Science 145:279.

Ortega, E. I., and Bates, L. S. 1983. Biochemical and agronomic studies of two modified hard-endosperm Opaque-2 maize (Zea mays L.) populations. Cereal Chem. 60:107.

Paulis, J. W., and Bietz, J. A. 1986. Separation of alcohol-soluble maize proteins by reversed-phase high performance liquid chromatography. J. Cereal Sci. 4:205.

Paulis, J. W., Bietz, J. A., Lambert, R. J., and Villegas, E. M. 1991. Identification of modified high-lysine maize genotypes by reversed-phase high-performance liquid chromatography. Cereal Chem. 68:361.

Pedersen, K., Argos, P., Naravana, S. V. L., and Larkins, B. A. 1986. Sequence analysis and characterization of a maize gene encoding a high-sulfur zein protein of M_r 15,000. J. Biol. Chem. 261:6279.

Prat, S., Perez-Grau, L., and Puigdomenech, P. 1987. Multiple variability in the sequence of a family of maize endosperm proteins. Gene 52:41.

Quality Protein Maize. 1988. National Academy Press, Washington, DC.

Robutti, J. L., Hoseney, R. C., and Wassom, C. E. 1974. Modified opaque-2 corn endosperms. II. Structure viewed with a scanning electron microscope. Cereal Chem. 51:173.

Subramanyam, M., Deyoe, C. W., and Harbers, L. H. 1980. Corn and sorghum. 1. Relationship of grain maturity to nutritional composition. Nutr. Rep. Int. 22:657.

Thompson, G. A., and Larkins, B. A. 1989. Structural elements regulating zein gene expression. BioEssays 10:108.

Vitale, A., Smaniotto, E., Longhi, R., and Galante, E. 1982. Reduced soluble proteins associated with maize endosperm protein bodies. J. Exp. Bot. 33:439.

Wallace, J. C., Lopes, M. A., Paiva, E., and Larkins, B. A. 1990. New methods for extraction and quantitation of zeins reveal a high content of γ-zein in modified opaque-2 maize. Plant Physiol. 92:191.

Wolf, M. J., Buzan, C. L., MacMasters, M. M., and Rist, C. E. 1952. Structure of the mature corn kernel. III. Microscopic structure of the endosperm of dent corn. Cereal Chem. 29:349.

Wolf, M. J., and Khoo, U. 1970. Mature cereal grain endosperm: Rapid glass knife sectioning for examination of proteins. Stain Technol. 45:277.

Wolf, M. J., Khoo, U., and Seckinger, H. L. 1967. Subcellular structure of endosperm protein in high-lysine and normal corn. Science 157:556.

Wolf, M. J., Khoo, U., and Seckinger, H. L. 1969. Distribution and subcellular structure of endosperm protein in varieties of ordinary and high-lysine maize. Cereal Chem. 46:253.

A GENETIC, BIOCHEMICAL AND ULTRASTRUCTURAL ANALYSIS
OF MODIFIED *opaque-2* MAIZE

Brian A. Larkins, Mauricio A. Lopes
Department of Plant Sciences, University of Arizona
Tucson, Arizona 85721

and Craig R. Lending
Department of Biological Sciences, SUNY Brockport
Brockport, New York   14420

INTRODUCTION

The search for maize genotypes with enhanced
nutritional quality led to the identification of
opaque-2 and *floury-2* as potentially useful mutations
for increasing the lysine content of the grain (Mertz
et al. 1944; Nelson et al., 1965). These mutations
reduce the synthesis of the most abundant seed
proteins, the storage proteins or zeins. Normally,
zeins account for more than half the endosperm
protein, and because they are devoid of the essential
amino acids lysine and tryptophan, maize seed is of
poor nutritional quality for monogastric animals.

The research initiated by Mertz and Nelson
resulted in the subsequent identification of several
genes that are believed to regulate storage protein
biosynthesis (see Motto et al. 1989 for a review), and
ultimately led to the isolation and characterization
of the genes encoding zeins. SDS-PAGE analysis of
zeins reveals a mixture of polypeptides, the most
predominant of which have molecular masses of Mr
27-kD, 22-kD, 19-kD, 16-kD, 14-kD, and 10-kD. The
unusual structure of these proteins causes them to
migrate faster or slower than their molecular weights

predict. It also causes them to dissolve in aqueous
/alcoholic solvents to varying extents. Both of these
factors are responsible for a complex nomenclature for
zein proteins (Esen, 1986; Hagen and Rubenstein,
1980). We recently suggested a classification system
based on protein structure (Larkins et al., 1989),
rather than differences in mobility on SDS-PAGE
(Wilson, 1991) or solubility (Esen, 1986). With this
system, the Mr 22-kD and Mr 19-kD proteins, which are
normally the most abundant, are called alpha-zeins,
the Mr 14-kD protein is called the beta-zein, the Mr
27-kD and 16-kD proteins are called gamma-zeins, and
the Mr 10-kD protein is called the delta-zein.

The alpha-zein proteins range in size from 210
to 245 amino acids. These proteins have high contents
of glutamine (25%), leucine (20%), alanine (15%), and
proline (11%), and, as is true of all the other zeins,
they contain no lysine (Shotwell and Larkins, 1989).
A distinguishing feature of these proteins is a series
of tandemly repeated peptides of approximately 20
amino acids that appear to form alpha-helices that
fold the protein into rod-shaped molecules (Argos et
al., 1982). The beta-zein protein is 160 amino acids
long and contains less glutamine (16%), leucine (10%),
and proline (9%), than the alpha-zeins, but has
significantly more methionine (4%) and cysteine (7%).
Unlike the alpha-zeins, the beta-zein contains no
alpha-helical structure and is composed mostly of
beta-strand and turn conformation (Pedersen et al.,
1986). The Mr 27-kD gamma-zein is a cysteine-rich
(7%) protein of 180 amino acids, with an exceptionally
high content of proline (25%) (Prat et al., 1987).
The high proline content is a consequence of a series
of hexapeptide repeats (PPPVHL) at the NH_2-terminus.
There are eight identical copies of this repeat in the
Mr 27-kD gamma-zein, but only three copies of a
homologous repeat in the Mr 16-kD gamma-zein. The
delta-zein is a small protein of 130 amino acids
(Kirihara et al., 1988). It is exceptionally rich in
the sulfur amino acids, methionine (23%) and cysteine
(4%).

Mutations like *opaque*-2 typically result in a
reduction in the synthesis of alpha-zeins. As a
consequence, endosperms of these mutants have a floury

rather than a hard, vitreous texture, and this causes
them to be opaque rather than translucent when placed
on a light box. Recent studies have shown that the
Opaque-2 gene encodes a leucine zipper type of
transcriptional regulatory protein that binds to the
promoters of alpha-zein genes and is required for
their transcription (Lohmer et al., 1991; Schmidt et
al., 1990). The *opaque*-2 mutants produce half as much
alpha-zein as their wild type counterparts, and this
leads to a two- to three-fold increase in the
percentage of lysine in the endosperm. Although
opaque-2 mutants were found to be nutritionally
superior to normal maize, the soft, floury endosperm
of the mutant results in lower kernel density and
increased susceptibility to insects, pathogens, and
mechanical damage. These problems, as well as the
reduced yield and protein content, prevented *opaque*-2
from becoming of commercial importance (Ortega and
Bates, 1983).

Some of the initial research with *opaque*-2
revealed that in certain genetic backgrounds the
floury phenotype of the mutant is modified such that

Fig. 1. Total Zein Proteins from Several Genotypes of
Maize. Protein extracts equivalent to 1 mg of
endosperm flour was subjected to SDS-PAGE and stained
with coomassie brilliant blue (as described by Wallace
et al., 1990). Lane a, extract from population
63-Blanco Dentado-1 QPM; b, population 64-Blanco
Dentado-2 QPM; c, population 62-White Flint QPM; d,
population 65-Yellow Flint QPM; e, population
66-Yellow Dent QPM; f, Pool 25 QPM; g, Pool 26 QPM; h,
Blanco Cristalino QPM; i, Amarillo Cristalino QPM; j,
La Posta QPM; k, Obregon 7940; l, Poza Rica 7940; m,
Guanacaste 7940; n, population 69-Templado Amarillo
QPM; o, population 70-Templado Amarillo QPM; p, Pool
33 QPM; q, Pool 34 QPM; r, Amarillo del Bajio QPM; s,
Amarillo Subtropical QPM; t, Templado Blaco Dentado
QPM; u, Obregon 7941; w, San Jeronimo 7941; x, Tuxpeno
normal; y, B73 normal; aa, W22 normal; bb, W64A
normal; cc, W64A*o2*; dd, W64A*fl2*; ee, W64A*wx*; ff, Oh43
normal; gg, Poza Rica *o2*. QPM germplasm is described
in CIMMYT (1987). (From Wallace et al., 1990, by
permission of ASPP)

TOTAL ZEIN EXTRACTIONS

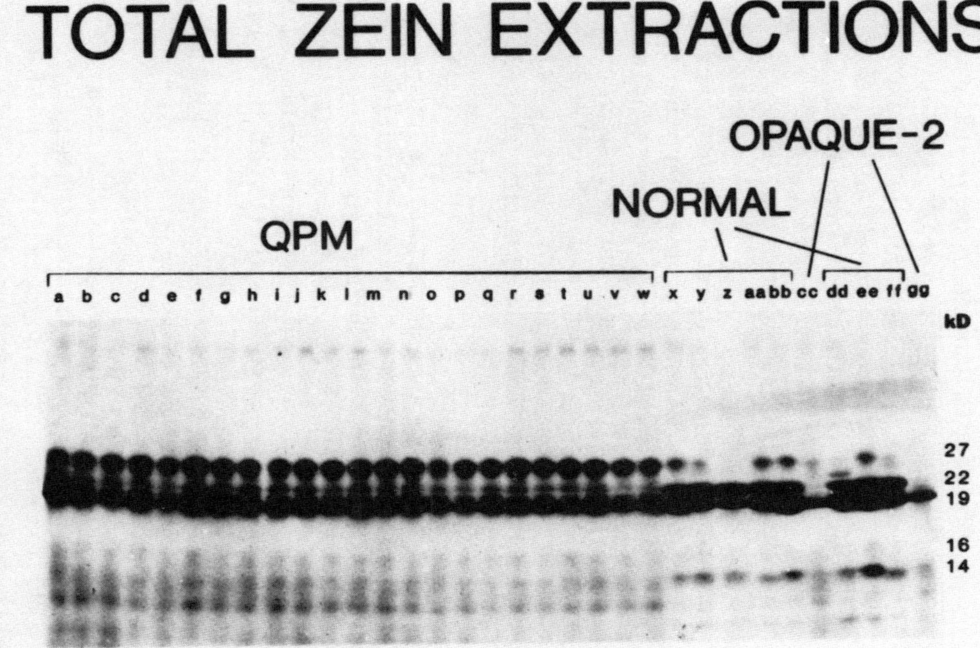

Figure 1

some regions of the endosperm are more vitreous than others (Paez, 1969). Genes that condition this phenotype are called "*opaque*-2 modifiers". The ability of these genes to convert soft endosperm *opaque*-2 mutants to a vitreous phenotype provided the basis for the development of modified *opaque*-2 mutants that have a normal protein concentration and a vitreous endosperm, as well as a lysine content that is at least double that of normal genotypes (Ortega and Bates, 1983). Through back-crossing and recurrent selection, breeders at the International Maize and Wheat Improvement Center (CIMMYT) generated a large number of these varieties, which are called Quality Protein Maize (QPM) (Vasal, 1980).

The mechanisms by which *opaque*-2 modifiers alter the phenotype of *opaque*-2 mutants has been the focus of our research for several years. Early studies showed that the conversion of a floury to a vitreous endosperm is associated with increased synthesis of a storage protein fraction (Gentinetta et al., 1975; Ortega and Bates, 1983). We subsequently showed that the principal change in storage protein is a two- to three-fold increase in gamma-zein synthesis (Wallace et al., 1990). To better characterize the changes associated with gamma-zein synthesis in modified *opaque*-2 kernels, we have characterized the synthesis of this protein in standard *opaque*-2 and modified *opaque*-2 genotypes and analyzed the inheritance of the gamma-zein content in reciprocal crosses of both types of mutants.

MATERIALS AND METHODS

Seeds of several mutant and normal maize inbred lines, QPM genotypes developed at CIMMYT-Mexico, their reciprocal (F1) crosses, F2 and backcross (BC1) progenies were obtained by hand pollination during the summers of 1988 and 1989. A CIMMYT report provides detailed description of the QPM germplasm used in this study (CIMMYT, 1987).

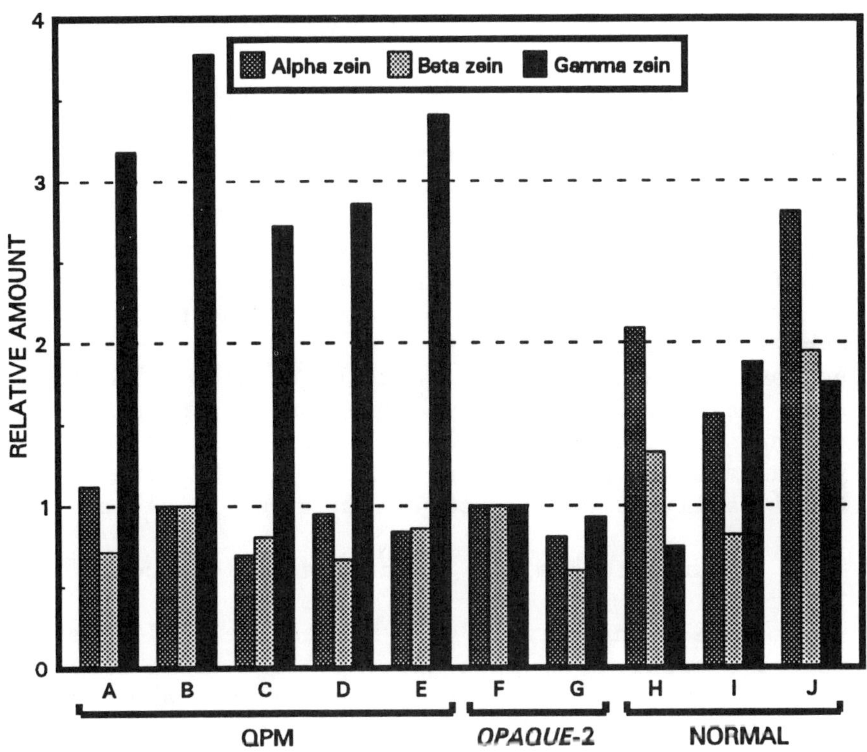

Fig. 2. Quantification of Alpha-, Beta-, and
Gamma-Zeins. ELISA assays were performed and analysed
as in Wallace et al., (1990). Amounts of protein were
normalized to that of the inbred line W64Ao2
(arbitrarily stablished as 1) (F). A, population
64-Blanco Dentado-2 QPM; B, population 65-Yellow Flint
QPM; C, population 66-Yellow Dent QPM; D, Pool 25 QPM;
E, Templado Blanco Dentado QPM; F, W64Ao2; G, Poza
Rica o2; H, B73 normal; I, W22 normal; J; W64A normal.
(From Wallace et al., 1990, by permission of ASPP)

Protein Analysis

Zeins were purified and antibodies specific for
alpha-, beta-, gamma-, and delta-zeins were prepared
and characterized as described by Lending et
al.(1988). Total zein proteins were extracted from
lyophilized powder of developing kernels as described

by Wallace et al. (1990). SDS-PAGE analyses were
performed in 7.5 to 18% polyacrylamide gradient gels
with the high molarity Tris buffer system of Fling and
Gregerson (1986). The relative concentration of zein
proteins was measured from duplicate samples using the
enzyme linked immunosorbent assay (ELISA) described by
Wallace et al. (1990). For all ELISA assays,
absorbance at 410 nm was measured over a series of
two-fold dilutions of the antigen, and a regression
analysis was performed on a range of concentrations
over which absorbance versus relative antigen
concentration approximated a straight line. In such
case, relative antigen concentration is proportional
to the slope of the regression line. Results were
normalized to those from a standard inbred line
(arbitrarily stablished as 1), included in all assays.

Immunocytochemical Staining and Microscopy

Developing kernels were harvested at different
stages (described as days after pollination - DAP),
immersed in fixative, embedded and immunostained for
analysis by light and electron microscopy as described
by Lending and Larkins (1989).

RESULTS AND DISCUSSION

The variation in zein composition of normal,
opaque-2, and QPM mutant varieties is illustrated in
Fig. 1. In the normal genotypes (lanes x, y, z, aa,
bb, ff) the 22-kD and 19-kD alpha-zeins are the most
abundant proteins. There is variation in the amount
of beta- and gamma-zeins, and typically only small
amounts of delta-zeins are present. A dramatic
reduction in 22-kD alpha-zeins, as well as a
significant reduction in 19-kD alpha-zeins is
characteristic of *opaque*-2 mutants (Fig. 1, lanes cc,
gg). Since QPM materials are also *opaque*-2 mutants, a
reduction in the 22-kD alpha-zeins is also evident in
these samples (Fig. 1, lanes a-w). However, the most
dramatic difference in the zein composition of QPM and
non-QPM genotypes is the amount of gamma-zein. This
protein is found in high concentrations in all of the
QPM materials, while it is present in small or

moderate amounts in the non-QPM genotypes.

Because quantification of zeins based on coomassie blue staining of SDS-gels is imprecise, we developed a procedure for measuring the amount of the various zein classes with antibodies using an ELISA assay. When suitable reaction conditions are used (Wallace et al., 1990), the enzyme coupled antibody reaction produces absorbance values that are directly proportional to the amount of protein (antigen) over at least a 10-fold range of concentrations.

Figure 2 shows a compilation of the ELISA results for several QPM, *opaque*-2, and normal maize varieties. To standardize the results, the protein quantities were normalized to those from W64A*o*2, which was included in each assay. These results are consistent with those obtained by SDS-PAGE (Fig. 1); modified *opaque*-2 mutants contain 2- to 3-times more gamma-zein than normal or *opaque*-2 genotypes. The amounts of alpha- and beta-zeins in QPM are similar to those in unmodified *opaque*-2, i.e. about one-half to one-third of normal.

Because the amount of gamma-zein was the most significant protein change in modified versus unmodified *opaque*-2 mutants, we investigated the inheritance of gamma-zein content in F1 progeny of both types of *opaque*-2 mutants, as well as in normal and *floury*-2 mutants. Since the endosperm is triploid, analysis of F1 hybrids from reciprocal crosses allowed us to compare the content of gamma-zein in genotypes theoretically containing zero, one, two, and three doses of modifier genes. Figure 3 shows typical kernel samples from several QPM genotypes and their reciprocal crosses with *opaque*-2 mutants. In contrast with the opaque types (Fig 3., column 4, A-H), QPM kernels (Fig. 3, column 1, A-K) have a vitreous, translucent endosperm. The parental lines and their reciprocal crosses (Fig. 3, columns 1 to 4, A-H) show a phenotypic gradient, with an increase of vitreousness toward the QPM types. Because *opaque*-2 is a recessive mutation, kernels from reciprocal crosses of the modified types by the normal inbred line W64A show only vitreous phenotypes (Fig. 3, columns 1 to 4, I-K).

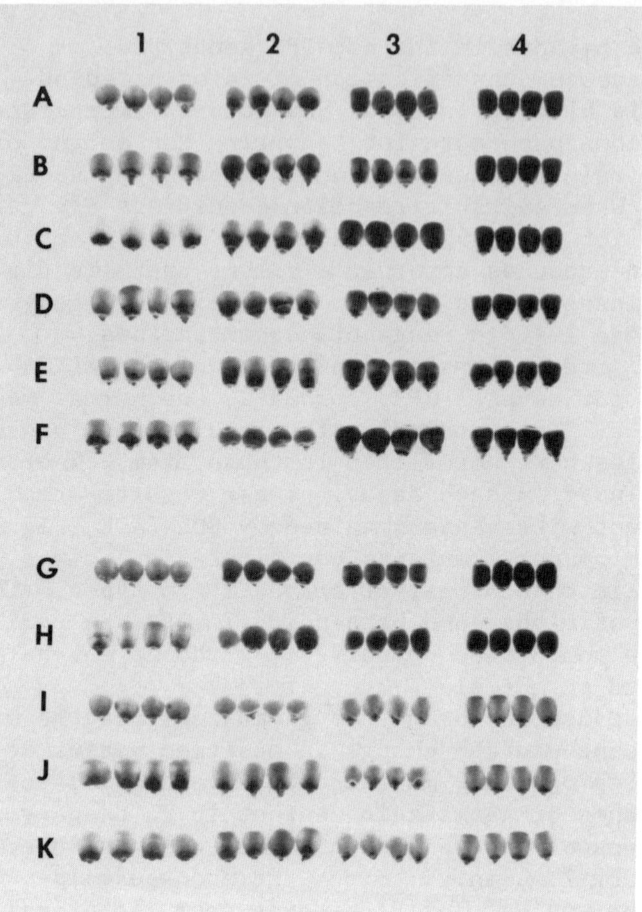

Fig. 3.   Photograph of Backlit Kernels Illustrating Opaque and Vitreous Phenotypes.   Columns 1 and 4 represent kernels from parent 1 (P1) and 2 (P2); columns 2 and 3 the reciprocal crosses P1 x P2 and P2 x P1, respectively.   A: Pool 25 QPM (P1) and W64A*o2* (P2);   B: Pool 34 QPM (P1) and W64A*o2* (P2);   C: Pool 23 QPM (P1) and W64A*o2* (P2);   D: Yellow Flint QPM (P1) and PA91*o2* (P2);   E: Pool 33 QPM (P1) and PA91*o2* (P2); F: Pool 34 QPM (P1) and OH43*o2* (P2);   G: Pool 25 QPM (P1) and W64A*o2f12* (P2);   H: Yellow Flint QPM (P1) and W64A*f12* (P2);   I: Pool 25 QPM (P1) and W64A+ (P2);   J: Pool 34 QPM (P1) and W64A+ (P2);   K: White Flint QPM (P1) and W64A+ (P2).   (From Lopes and Larkins, 1991, by permission of Crop Science Society of America, Inc.)

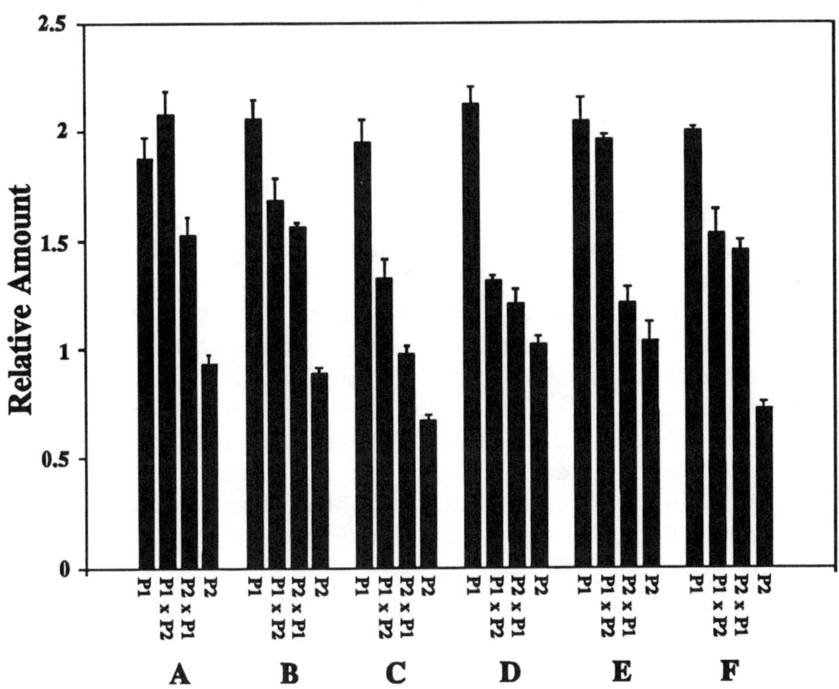

Fig. 4. Quantification by ELISA of Gamma-Zein
Extracted from Several QPM, Unmodified *o2*, and their
Reciprocal Crosses. ELISA assays were performed and
analysed as in Wallace et al., (1990). Amounts of
protein were normalized to that of the inbred line
W64A (arbitrarily stablished as 1), which was included
in all assays. A: Pool 25 QPM (P1), and W64A*o2* (P2);
B: Pool 34 QPM (P1), and W64A*o2* (P2); C: Pool 34 QPM
(P1), and Oh43*o2* (P2); D: Yellow Flint QPM (P1), and
PA91*o2* (P2); E: Pool 33 QPM (P1), and PA91*o2* (P2); F:
Pool 23 QPM (P2), and Oh43*o2* (P2). Lines on the top
of the bars represent the standard deviations. (From
Lopes and Larkins, 1991, by permission of Crop Science
Society of America, Inc.)

Figure 4 shows the measurement of gamma-zein
from some of the parental and reciprocal F1 hybrids
ilustrated in Figure 3. The ELISA assays show that
the QPM genotypes have about two times as much
gamma-zein as unmodified *opaque-2* lines, and the
relative amounts for the F1 reciprocal crosses

increased gradually toward the QPM parent.

To further examine endosperm modification, we
characterized kernel vitreousness in segregating
progenies of crosses between QPM and unmodified
opaque-2 inbred lines. Figure 5 shows kernels
selected from individual, well-developed ears of F2
and first backcross progenies. The segregation
patterns for kernel vitreousness are typical of the
genotypes shown. For example, B37*o2* by Pool 33 QPM
(Fig. 5E) seggregating kernels have an opaque
phenotype varying from the tip to the crown. The F2
progeny of Co22*o2* by White Flint QPM (F) show opaque
patches irregularly interspersed within vitreous
regions. One back cross with unmodifed *opaque*-2 as
the recurrent parent results in a substantial
reduction of vitreousness, as shown for progenies of
(W64A*o2* X Pool 34 QPM, F1) by W64A*o2* (J) and (W64A*o2* X
Blanco Dentado QPM, F1) by W64A*o2* (L).

The content of gamma-zein in the F2 progeny
co-seggregated with the vitreous phenotype. This is
illustrated in Figure 6, which shows the ELISA
analysis of gamma-zein content in endosperms from a
segregating ear of a White Flint QPM by Co22*o2* cross.
Kernels from a single well-developed ear were
separated into three distinct phenotypic classes:
opaque, semi-opaque, and vitreous. Three samples of
15 kernels from each class were analyzed along with
samples of the modified and unmodified parents. The
results of this analysis show that the QPM type (Fig.
6E) has more than 2.5 times the content of gamma-zein
as the unmodified parent (Fig. 6A). The opaque and
normal segregating types (Fig. 6B and 6D) do not
contain the same amount of gamma-zein as the opaque
and modified parents, probably as a consequence of the
quantitative nature of endosperm modification, which
makes it difficult to select totally opaque and
totally modified types from a single segregating ear.
However, there is a trend toward a parallel increase
in gamma-zein protein and vitreous phenotype.

Zein proteins are synthesized on rough
endoplasmic reticulum (RER) membranes, and they
aggregate into protein bodies within the lumen of the
RER (Larkins and Hurkman, 1978). Based on
immunocytochemical staining of protein bodies in

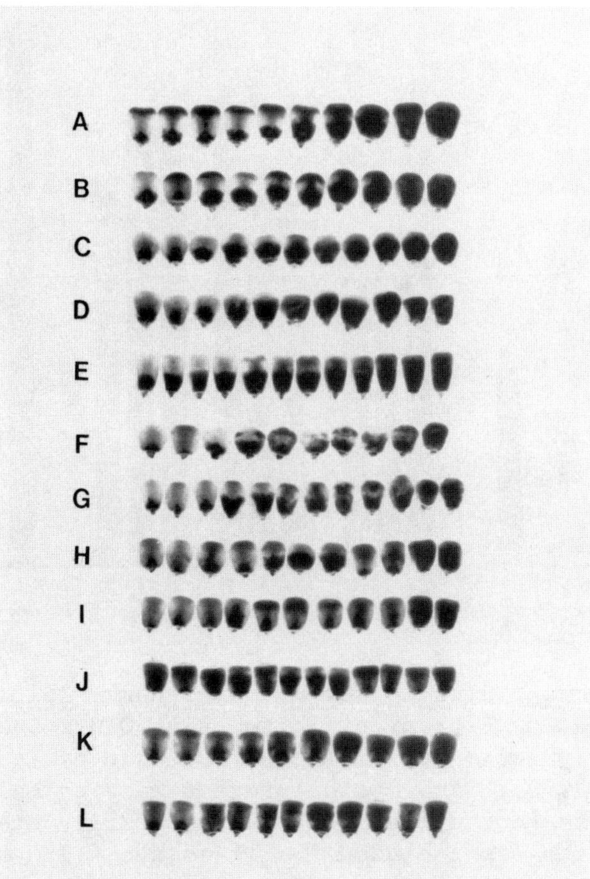

Fig. 5. Photograph of Backlit Kernels from Segregating Progenies of QPM (Modified) x Unmodified *o2* Maize. Kernel phenotype ranges from vitreous (left) to opaque (right). A: Mo17*o2* x Pool 18 QPM, F2; B: Mo17*o2* x Pool 23 QPM, F2; C: Oh43*o2* x Blanco Dentado QPM, F2; D: Oh42*o2* x Pool 33 QPM, F2; E: B73*o2* x Pool 33 QPM, F2; F: Co22*o2* x White Flint QPM, F2 G: W117*o2* x White Flint QPM, F2; H: W64A*o2* x Pool 23 QPM; I: W64A*o2* x Pool 34 QPM, F2; J: (W64A*o2* x Pool 34 QPM, F1) x W64A*o2*, BC1; K: W64A*o2* x Blanco Dentado QPM, F2; L: (W64A*o2* x Blanco Dentado QPM, F1) x W64A*o2*, BC1. (From Lopes and Larkins, 1991, by permission of Crop Science Society of America, Inc.)

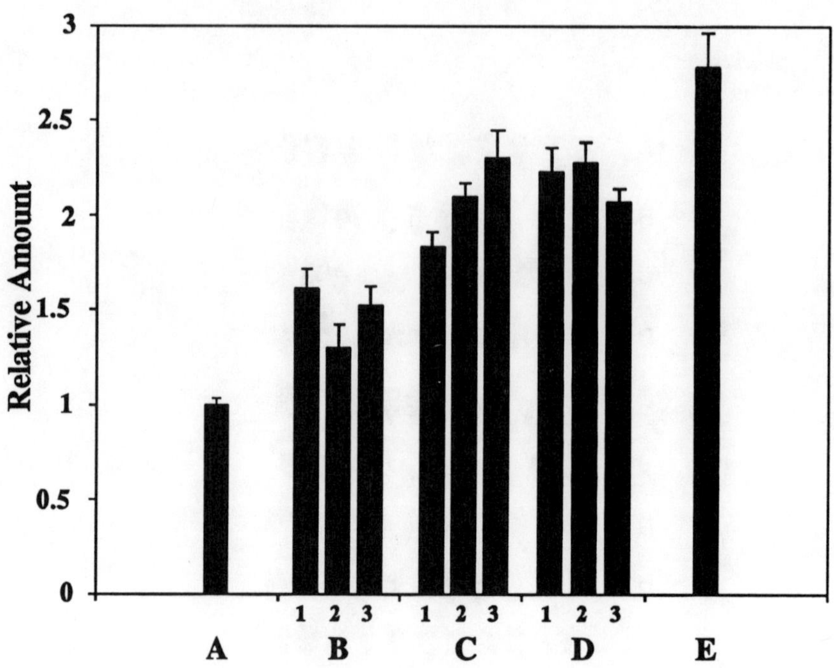

Fig. 6.   Quantification by ELISA of Gamma-Zein from an
F2 Segregating Progeny of White Flint QPM x Co22o2.
ELISA assays were performed and analysed as in Wallace
et al., (1990).   Amounts of protein were normalized to
that of the inbred line Co22o2 (arbitrarily stablished
as 1), which was included in all assays.   B,C and D
represent triplicates of opaque, semi-opaque and
vitreous samples of an individual segregating ear.
White Flint (E) is the QPM parent.   Lines on the top
of the bars represent the standard deviations.   (From
Lopes and Larkins, 1991, by permission of Crop Science
Society of America, Inc.)

---

developing endosperm, protein body formation appears
to involve an orderly association of specific classes
of zeins (Lending and Larkins, 1989).   Figure 7A shows
a section of normal endosperm at 18 days after
pollination (DAP) stained with basic fuchsin.   The
outer cell layer, the aleurone, contains many lipid
bodies and is cytologically distinct from the deeper
endosperm cells.   The endosperm cells become larger

with increasing distance from the aleurone layer.
Protein bodies and starch grains are the predominant
components in these cells. The size of the protein
bodies gradually increases with the distance from the
first subaleurone layer; those in the sixth starchy
endosperm layer are approximately 1 μm in diameter.
Immunocytochemical staining revealed that this change
in protein body size is associated with the appearance
of different types of zeins.

In sections immunolabeled with alpha-zein
antibodies, light staining is observed in the
subalerone cell layer and in the first layer of
starchy endosperm, although it is not apparent at the
magnification shown in Figure 7B. The intensity of
staining increases in deeper cell layers of the
endosperm and becomes constant by the third starchy
endosperm cell layer. The immunostaining forms solid
circular deposits; the diameter of the deposits
increases from the third subaleurone layer inward.
The staining patterns are similar in size to that of
the underlying protein bodies (Fig. 7B, inset).

Immunostaining with beta-zein antibodies is
visible within the first subaleurone layer (Fig. 7C),
unlike that observed for the alpha-zein. The labeling
is most intense in the second through the fourth
starchy endosperm layers and becomes reduced in cells
farther from the aleurone. The staining in the first
few cells beneath the aleurone layer formed solid
circular patterns (not apparent at this
magnification). In cells farther from the aleurone
layer, the immunolabeling for beta-zein forms open
circular patterns of varying intensity. While most of
the staining occurred over the peripheral regions of
the protein bodies, some central inclusions often
stained (Fig. 7C, inset).

The pattern of staining with the gamma-zein
antibody is similar to that for beta-zein; labeling is
observed in the first few cell layers immediately
beneath the aleurone (Fig. 7D) and decreases at
greater distances from the aleurone layer. The
staining in the first few cells beneath the aleurone
forms solid circular patterns, although it is not as
dense at that observed for the beta-zein (not apparent
at the magnification shown). Although the labeling

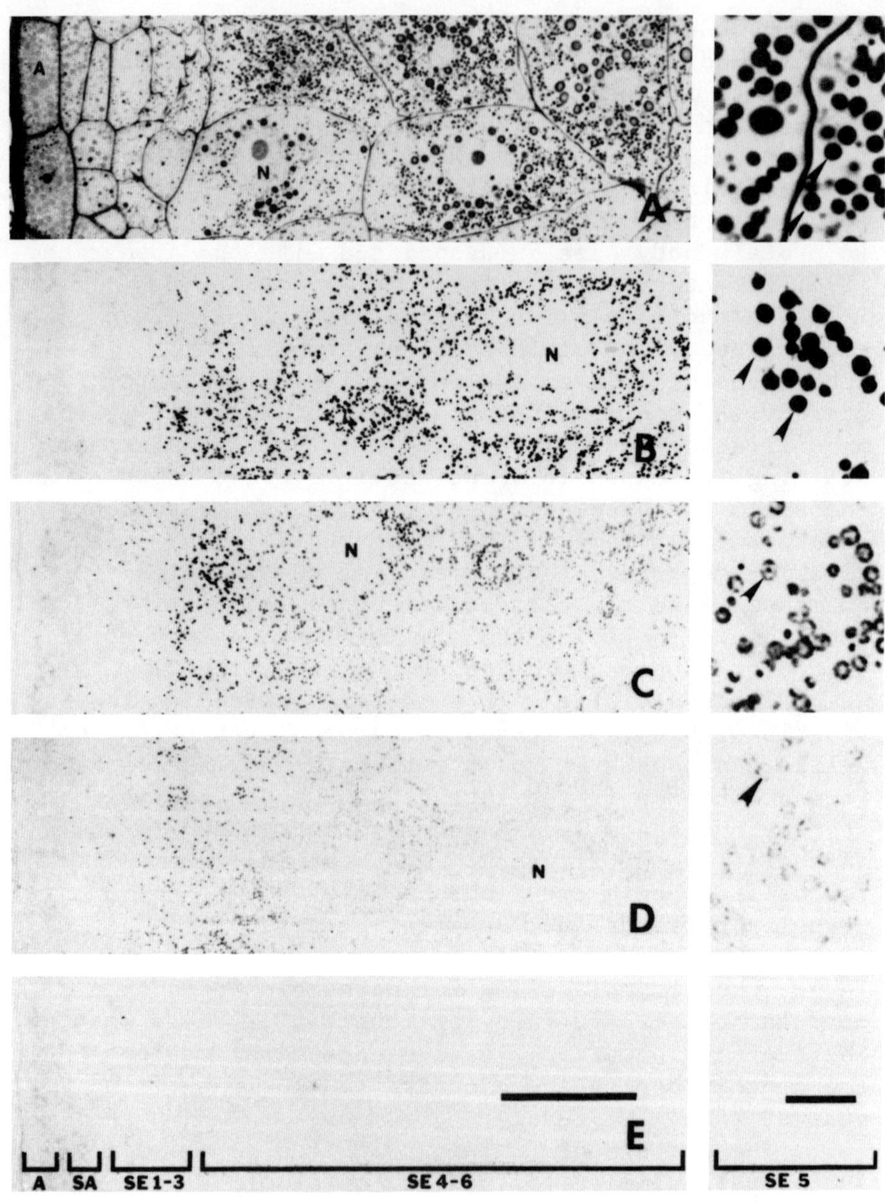

Figure 7

Fig. 7. Photomicrographs of Endosperm at 18 DAP,
Illustrating the Immunostaining Patterns with
Antibodies Directed Against Alpha-, Beta-, and
Gamma-zeins. Bar = 50 μm, and bar = 5 μm for all
insets. Brackets at the bottom indicate approximate
cell position. A, aleurone; SA, subaleurone; SE,
starchy endosperm. (A) Sections stained with basic
fuchsin illustrating a region similar to those
immunostained with the various antibodies. (Inset)
High magnification view of the fifth starchy endosperm
layer. Most protein bodies from this region of the
endosperm are from 0.8 to 1.4 μm in diameter
(arrowheads).
(B, C, and D) Sections incubated with alpha-, beta-,
and gamma-zein antibodies respectively, followed by
goat anti-rabbit/coloidal gold, and enhanced with
silver intensification reagent. (Insets) High
magnification view of the fifth starchy endosperm
layer. See detailed description in the text.
(E) Section immunostained with preimmune serum,
followed by goat anti-rabbit/colloidal gold, and
enhanced with silver intensification reagent. (From
Lending and Larkins, 1989, by permission of ASPP).

---

pattern for gamma-zein in the interior cells was
similar to the beta-zein, the intensity of staining is
less for the gamma-zein (Fig 7D., inset).

Figure 8 shows a series of electron micrographs
illustrating changes in the morphology and size of
protein bodies in the subaleurone, first, second, and
fifth starchy endosperm cell layers at 14 DAP.
Protein bodies in the subalurone layer are between 0.1
and 0.3 μm in diameter and consist primarily of
dark-staining material; infrequently light-staining
deposits are observed. Protein bodies in the first
starchy endosperm layer range in size from 0.2 to 0.5
μm in diameter, and consist mostly of dark-staining
material with locules of light-staining material
toward the outer regions. Protein bodies in the
second starchy endosperm layer range from 0.3 to 0.6
μm and consist mostly of a central core of
light-staining material surrounded by a peripheral
ring of dark-staining material. Small deposits of

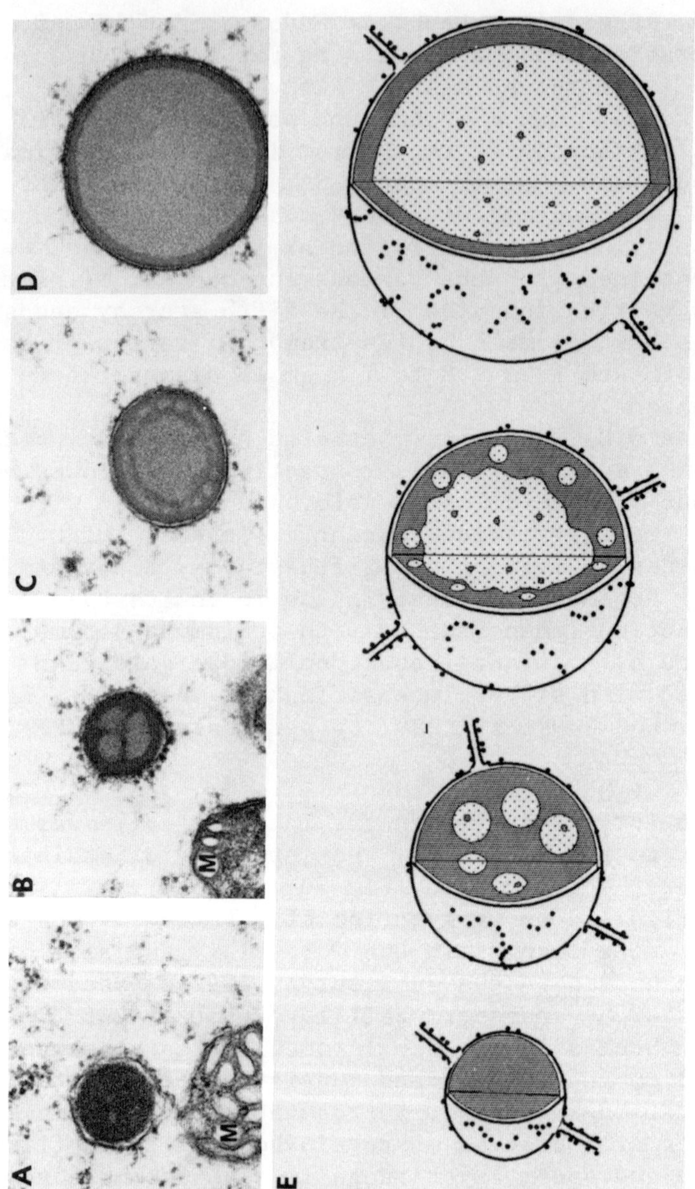
A
B
M
C
D
M
E
Figure 8

Fig. 8. (A to D) Electron Micrographs Showing
Representative Protein Bodies from Different Cell
Layers of the Endosperm at 14 DAP. The protein bodies
are representative of the morphology and size of the
protein bodies within the cell layer indicated, and
represent a developmental series discussed in the text
(from left to right). The sections were stained with
uranyl acetate and lead citrate. (A) Protein bodies
from the subaleurone layer. (B) Protein bodies from
the first starchy endosperm layer. (C) Protein
bodies from the second starchy endosperm layer. (D)
Protein bodies from the fifth starchy endosperm layer.
(E) A model for the pattern of zein deposition during
protein body formation based on the above examination
of the distribution of the various zeins in protein
bodies. The heavily stippled regions correspond to
regions that are rich in beta- and gamma-zeins, and
the lightly stippled regions correspond to regions
rich in alpha-zein. The protein body is surrounded by
rough endoplasmic reticulum (dark dots represent
ribosomes). Some beta- and gamma-zeins are found
within the regions that consist primarily of
alpha-zein (heavily stippled inclusions). See text
for details. (From Lending and Larkins, 1989, by
permission of ASPP)

---

dark-staining material are frequently observed within
the central region (not shown).

Immunogold labeling of similar sections revealed
that the light-staining regions in the protein bodies
corresponded to alpha-zeins, while the dark-staining
region contained both beta- and gamma-zeins (Lending
and Larkins, 1989). We have subsequently determined
that the delta-zein also occurs in the light-staining
central region (Lending and Larkins, unpublished).

These changes in the protein composition and
morphology of protein bodies in progressively deeper
cell layers of the endosperm are consistent the
developmental pattern of protein body formation
illustrated in the lower portion of Fig. 8. The
initial accretions within the RER consist of beta- and
gamma-zeins, which contain little or no alpha-zein.
Subsequently, alpha-zein begins to accumulate and is
observed as discrete, light-staining locules within

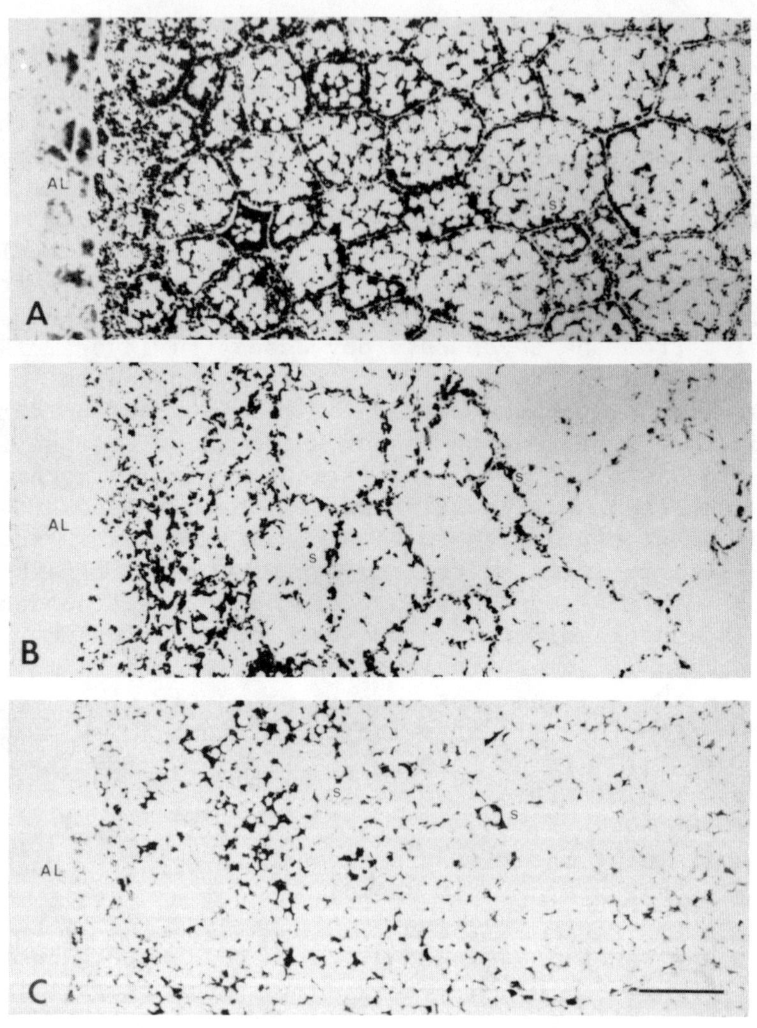

Figure 9

Fig. 9. Photomicrographs of Sections from Developing
Maize Endosperm at 30 DAP Illustrating the
Immunostaining Pattern with Alpha-Zein Antibodies.
Abbreviations: AL, aleurone; S, starch granule.
Sections were incubated with alpha-zein antibody,
followed by goat anti-rabbit/colloidal gold, and
visualized by enhancement with silver intensification
reagent (Lending and Larkins 1989). (A) W64A.
Although staining deposits are absent from the
aleurone, they are uniform throughout the endosperm
and surround starch granules. (B) W64A*o2*. Staining
deposits occur throughout the endosperm, but the level
of immunostaining is less than that of W64A. Also,
immunostaining tends to be near the peripheral region
of cells, adjacent to cell walls. (C) Pool 34 QPM.
Staining deposits are uniform throughout the endosperm
and surround starch granules. However, the level of
immunostaining is considerably reduced compared to
W64A. Bar = 50 μM. (From Geetha et al., 1991, by
permission of ASPP)

the beta- and gamma-zeins. As the interior of the
protein body fills with alpha-zeins, the locules of
alpha-zeins fuse and aggregate to form a central core;
some smaller locules of alpha-zein remain and are
interspersed in the outer region of the protein body.
The dark-staining region that contains beta- and
gamma-zein forms a continuous layer around the
periphery of the protein body, although there are
patches of beta-zein and, more commonly, gamma-zein
within the central region.

Based on the pattern of zein accumulation in
protein bodies of the normal genotype, one would
predict that the reduction in alpha-zein synthesis in
opaque-2 mutants and the increase of gamma-zein
synthesis in modified *opaque*-2 mutants would have a
profound affect on protein body formation. Figure 9
shows a light microscopy analysis of endosperm
sections from W64A, W64Ao2, and Pool 34 QPM that were
immunostained with alpha-zein antibodies. In sections
of endosperm sampled from the top of the seed at 30
DAP, immunolabeling with the alpha-zein antibody is
spatially similar among the three different genotypes.
In W64A (Fig. 9A), staining deposits are absent from

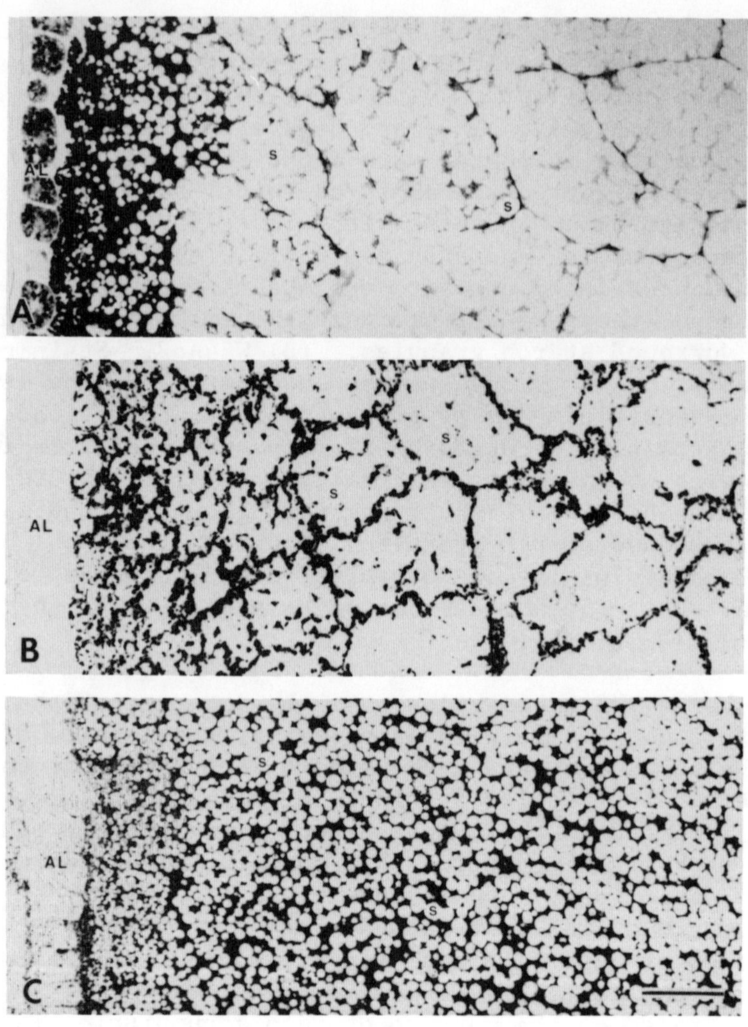

Figure 10

Fig. 10. Photomicrographs of Sections Through
Developing Maize Endosperm at 30 DAP Illustrating the
Immunostaining Pattern of Antibodies Against
Gamma-Zein. Abbreviations as in Figure 9. All
sections were incubated with gamma-zein antibody
followed by goat anti-rabbit/colloidal gold with
enhancement by silver intensification reagent. (A)
W64A. Most of the gamma-zein is in the layers of
cells adjacent to the aleurone and surrounds the
starch grains. Relatively little gamma-zein is
observed in cells farther away from the aleurone
layer. (B) W64A*o2*. Staining deposits occur
throughout the endosperm, predominantly in regions
near the periphery of cells. (C) Pool 34 QPM.
Staining deposits are uniform throughout the endosperm
and around the starch granules. The level of
immunostaining is considerably greater and more
uniform throughout the endosperm than in W64A or
W64A*o2*. Bar = 50 μM. (From Geetha et al, 1991, by
permission of ASPP)

---

the aleurone, but they are uniform throughout the
endosperm and surrounded the starch granules.
Considerably more immunostaining is observed for W64A
than W64A*o2* (Fig. 4B) or Pool 34 QPM (Fig. 4C).
In W64A*o2*, immunostaining tended to be near the
peripheral region of cells, adjacent to cell walls.
In Pool 34 QPM, staining deposits are uniform
throughout the endosperm and surrounded starch
granules. The variation in the intensity of
immunostaining reflected the difference in the
concentration of alpha-zeins in the developing seeds.

Comparable sections stained with antibodies
against gamma-zein are shown in Figure 10. In
sections of W64A endosperm sampled from the top of the
seed at 30 DAP (Fig. 10A), immunolabeling with the
gamma-zein antibody is most intense in the layers of
cells adjacent to the aleurone, and surrounded the
starch grains. Little gamma-zein was detected in
cells farther away from the first several subaleurone
cell layers. The gamma-zein distribution in W64A*o2*
(Fig. 10B) was similar to that in its normal
counterpart. The staining is most pronounced in the
first subaleurone cell layers. In more centrally

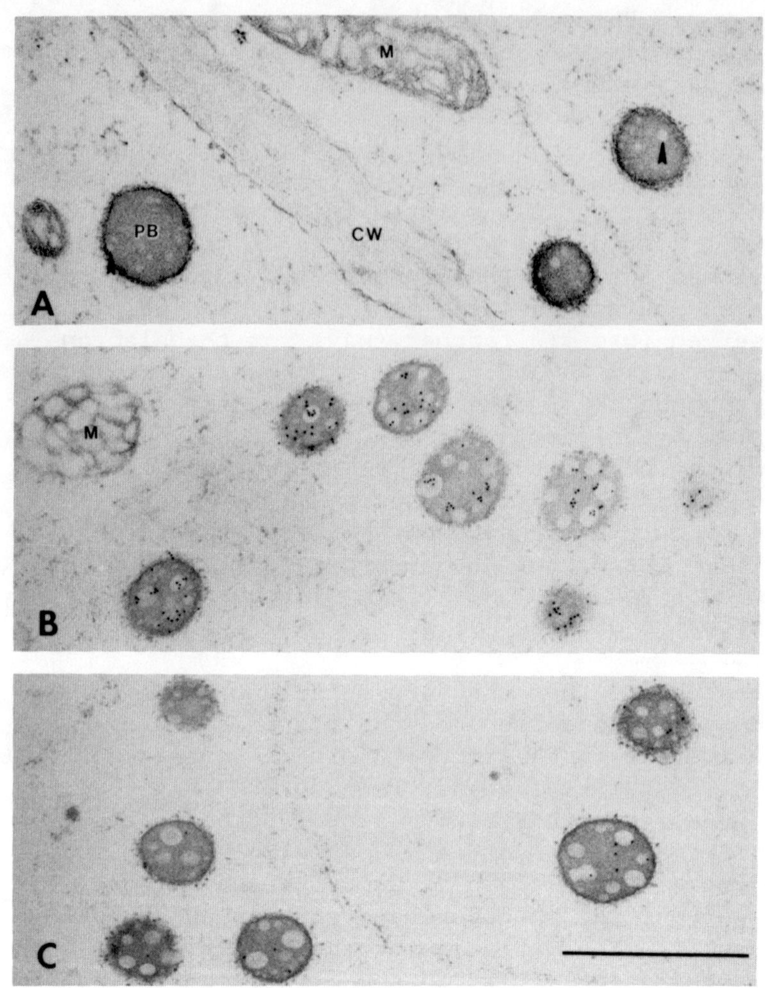

Figure 11

Fig. 11. Electron Micrographs Showing Representative
Protein Bodies of W64A*o2* at 22 DAP. Sections are from
endosperm tissue approximately 5 cells beneath the
aleurone layer. For immunostaining, sections were
incubated with the primary antibody and then with goat
anti-rabbit/colloidal gold (10 nm). Abbreviations: M,
mitochondrion; CW, cell wall; S, starch grain. (A)
Section poststained with uranyl acetate and lead
citrate. (B) Section immunostained for alpha-zein.
(C) Section immunostained for gamma-zein. Bar = 0.5
μM. (From Geetha et al., 1991, by permission of ASPP)

located cells, the most intense reaction is associated
with protein bodies near the cell wall.

Figure 10C shows the immunostaining of Pool 34
QPM sampled from the top of the seed at 30 DAP. The
amount and distribution of gamma-zein in these cells
are strikingly different from either of the W64A
genotypes. In Pool 34 QPM, not only is there intense
staining of gamma-zein in the cells just beneath the
aleurone layer, but it extends into the central region
of the endosperm. The intensity of the reaction is
even throughout these cells, such that the starch
grains are clearly highlighted and the cell walls are
not clearly demarked.

To determine the effect of the reduction of
alpha-zeins and increased content of gamma-zein in
modified *opaque*-2 mutants, we examined endosperm cells
of W64A*o2* and Pool 34 QPM by immunogold labeling and
electron microscopy. Figure 11 shows that the protein
bodies in W64A*o2* show light and dark staining regions,
similar to those observed in the subaleurone endosperm
cells of wild-type W64A (Fig. 8). However, few of the
protein bodies consist of a peripheral dark-staining
region surrounding the light-staining core that is
characteristic of protein bodies from the region
several cell layers beneath the aleurone in W64A. The
light-staining areas form locules within the
dark-staining region, but they seldom fuse into a
light-staining core. As shown in Figs. 11B and 11C,
immunogold labeling of W64A protein bodies with the
alpha-zein antibody occurrs predominantly over the
light-staining areas, whereas the gamma-zein

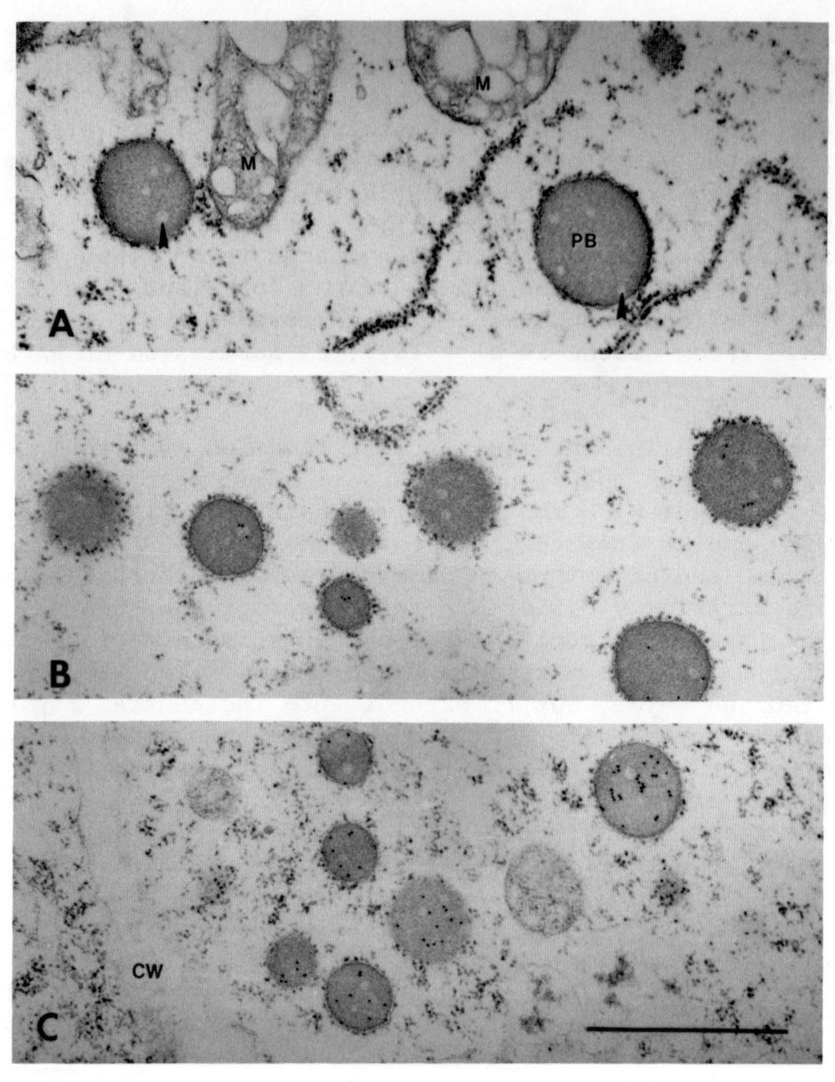

Figure 12

Fig. 12. Electron Micrographs Showing Representative
Protein Bodies of Pool 34 QPM at 22 DAP. Sections
from endosperm cell layer approximately 5 cells from
the aleurone layer. For immunostaining, sections were
treated as described in Figure 11. Bar = 0.5 μM.
Abbreviations as in Figure 11. (A) Section
poststained with uranyl acetate and lead citrate. (B)
Section immunostained for alpha-zein. (C) Section
immunostained for gamma-zein. (From Geetha et al.
1991, by permission of ASPP)

---

immunolabeling occurred over the dark-staining areas.
     Figure 12 shows sections of endosperm tissue
from Pool 34 QPM that are comparable to those of
W64Ao2 in Figure 11. The protein bodies of Pool 34
QPM are of similar size to those in W64Ao2 (Fig. 12A).
They stain darkly and contain small, light-staining
locules. Immunogold labeling with alpha-zein
antibodies (Fig. 12B) is sparse and occurs over the
light-staining areas. The immunogold labeling with
gamma-zein antibodies (Fig. 12C) occurs over the
dark-staining area and is much more intense than that
detected in W64Ao2.

DISCUSSION

     The results of these studies reveal that the
most significant change in maize endosperm as a
consequence of *opaque*-2 modifiers is the increased
synthesis of the gamma-zein protein. The content of
this protein is inherited in a codominant fashion,
depending on the dosage of modifier genes, and the
content of gamma-zein seggregates in F2 progeny along
with the vitreous characteristics of the kernel.
While these data do not prove that gamma-zein
synthesis is a consequence of modifier gene activity,
our results are consistent with the conclusion that
the two are very closely linked.
     It is highly plausible that the increased
synthesis of gamma-zein is a consequence of modifier
gene activity. The stocks used in QPM development
were obtained by selecting in an *opaque*-2 background
for seeds with a high protein content, a vitreous

phenotype, and a high lysine content. The *opaque*-2 gene encodes a transcription factor that is required for expression of many of the alpha-zein genes (Kodrzycki et al., 1989; Schmidt et al., 1990). However, transcription of the gamma-zein genes is not significantly affected by this mutation (Kodrzycki et al., 1989). Consequently, this selection scheme favored factors that increase the expression of the gamma-zein genes.

The mechanism by which an increase in gamma-zein content would result in a vitreous endosperm phenotype is unclear. It could result simply from a quantitive increase in protein, or it could be a qualitative effect of the gamma-zein protein itself. The hard, translucent phenotype could result from a higher concentration of protein in the cell, which results in less air space when the endosperm become desiccated. Alternatively, it could be related to the number and composition of protein bodies. The gamma-zein occurs near the periphery of the protein body where it is highly crosslinked by disulfide bonds (Lopes and Larkins, 1991). The majority of protein bodies in normal maize endosperms are 1-2 μm in diameter (Lending et al., 1988), whereas those from QPM and unmodified *opaque*-2 types are approximately 0.4 μm in diameter (Geetha et al., 1991). To account for the higher content of gamma-zein with reduced protein body size, the cytoplasm of QPM endosperm cells must have many more of these organelles than unmodified *opaque*-2 types. As the seed matures, the desiccation of the endosperm may lead to disintegration of RER membranes exposing the gamma-zein protein to the cytoplasmic matrix. During desiccation, the protein bodies become closely packed (Torrent et al., 1989) and possibly cross-linked by the gamma-zein protein, which could result in increased kernel vitreousness.

Regardless of the mechanism by which the increase in gamma-zein synthesis may cause the formation of a more vitreous kernel, the higher content of this protein in QPM genotypes cannot explain the high lysine content of the grain. The gamma-zein protein contains no lysine. The increased percentage of lysine in standard *opaque*-2 mutants is largely a consequence of the significant reduction in

alpha-zein proteins, which also contain no lysine.
The fact that QPM genotypes have a two- to three-fold
increase of gamma-zein content, while maintaining a
high lysine content implies that there must be a
significant increase in some lysine-rich protein
fraction in the grain. If the origin of this protein
fraction can be determined, potentially more
lysine-rich versions of QPM materials could be
developed.

LITERATURE CITED

Argos, P., Pedersen, K., Marks, D.M., and Larkins,
     B.A. (1982). A structural model for maize zein
     proteins. J. Biol. Chem. 257, 9984-9990.
CIMMYT. (1987). CIMMYT Report on Maize Improvement,
     1982-1983. Mexico, D.F.
Esen, A. (1986). Separation of alcohol-soluble
     proteins (zeins) from maize into three fractions
     by differential solubility. Plant Physiol.
     80:623-627.
Fling, S.P., and Gregerson, D.S. (1986). Peptide and
     protein molecular weight determination by
     electrophoresis using a high-molarity tris
     buffer system without urea. Anal. Biochem.
     155:83-88.
Geetha, K.B., Lending, C.R., Lopes, M.A., Wallace,
     J.C., and Larkins, B.A. (1991). *opaque*-2
     modifiers increase gamma-zein synthesis and
     alter its spatial distribution in maize
     endosperm. Plant Cell 3, 1207-1219.
Gentinetta, E., Maggiore, F., and Salamini, F. (1975).
     Protein studies in 46 *opaque*-2 strains with
     modified endosperm texture. Maydica 20, 145-164.
Graham, G.G., Lembcke, J., and Morales, E. (1990).
     Quality Protein Maize as the sole source of
     dietary protein and fat for rapidly growing
     young children. Pediatrics 85, 85-91.
Hagen, G. and Rubenstein, I. (1980). Complex
     organization of zein genes in maize. Gene 13,
     239-249.

Kirihara, J.A., Petri, J.B. and Messing, J. (1988).
    Isolation and sequence of a gene encoding a
    methionine-rich 10-kD zein protein from maize.
    Gene 71, 359-370.

Kodrzycki, R., Boston, R.S., and Larkins, B.A. (1989).
    The *opaque*-2 mutation of maize differentially
    reduces zein gene transcription. Plant Cell 1,
    105-114.

Larkins, B.A., and Hurkman, W.J. (1978). Synthesis and
    deposition of zein in protein bodies of maize
    endosperm. Plant Physiol. 62, 256-263.

Larkins, B.A., Lending, C.R., Wallace, J.C., Galili,
    G., Kawata, E.E., Geetha, K.B., Kriz, A.L.,
    Martin, D.N., and Bracker, C.E. (1989). Zein
    gene expression during maize endosperm
    development. p. 109-120. *In* R.B. Goldberg (ed.)
    The Molecular Basis of Plant Development, , Alan
    R. Liss, Inc., New York, N.Y.

Lending, C.R., Kriz, A.L., Bracker, C.E., and Larkins,
    B.A. (1988). Structure of maize protein bodies
    and immunocytochemical localization of zeins.
    Protoplasma 143, 51-62.

Lending, C.R., and Larkins, B.A. (1989). Changes in
    the zein composition of protein bodies during
    maize endosperm development. Plant Cell 1,
    1011-1023.

Lohmer, S., Maddaloni, M., Motto, M., Di Fonzo, N.,
    Hartings, H., Salamini, F., and Thompson, R.D.
    (1991). The maize regulatory locus Opaque-2
    encodes a DNA-binding protein which activates
    transcription of the b-32 gene. The EMBO J. 10,
    617-624.

Lopes, M.A., and Larkins, B.A. (1991). Gamma-zein
    content is related to endosperm modification in
    Quality Protein Maize. Crop Sci: 31, 1655-1662.

Mertz, E.T., Bates, L.S., and Nelson, O.E. (1964).
    Mutant gene that changes protein composition and
    increases lysine content of maize endosperm.
    Science 145, 279-280.

Motto, M., Di Fonzo, N., Hartings, H., Maddaloni, M.,
    Salamini, F., Soave, C., and Thompson, R.D.
    (1989). Regulatory genes affecting maize
    storage proteins. Oxford Surveys of Plant Mol.
    Biol. and Cell Biol. 6, 87-114.

Nelson, O.E., Mertz, E.T., and Bates, L.S. 1965. Second mutant gene affecting the amino acid pattern of maize endosperm proteins. Science 150, 1469-1470.

Ortega, E.I., and Bates, L.S. (1983). Biochemical and agronomic studies of two modified hard-endosperm *opaque*-2 maize (*Zea mays* L.) populations. Cereal Chem. 60, 107-111.

Paez, A.V., Helm, J.L., and Zuber, M.S. (1969). Lysine content of *opaque*-2 maize kernels having different phenotypes. Crop Sci. 9, 251-252.

Pedersen, K., Argos, P., Narayana, S.L.V., and Larkins, B.A. (1986). Sequence analysis and characterization of a maize gene encoding a high-sulphur zein protein of Mr 15,000. J. Biol. Chem. 261, 6279-6284.

Prat, S., Cortadas, J., Puigdomenech, P., and Palau, J. (1985). Nucleic acid (cDNA) and amino acid sequences of the maize endosperm protein glutelin-2. Nucleic Acids Res. 13, 1493-1504.

Schmidt, R.J., Burr, F.A., Aukerman, M.J., and Burr, B. (1990). Maize regulatory gene *opaque*-2 encodes a protein with a "leucine-zipper" motif that binds to zein DNA. Proc. Natl. Acad. Sci., USA 87, 46-50.

Shotwell, M.A., and Larkins, B.A. (1989). The molecular biology and biochemistry of seed storage proteins. *In* The Biochemistry of Plants: A Comprehensive Treatise. (A. Marcus, ed.) Vol. 15, pp. 297-345., Academic Press, San Diego, CA.

Torrent, M., Geli, M.I., and Ludevid, M.D. (1989). Storage-protein hydrolysis and protein-body breakdown in germinated *Zea mays* L. seeds. Planta 180, 90-95.

Vasal, S.K., Villegas, E., Bjarnason, M., Gelaw, B., and Goertz, P. (1980). Genetic modifiers and breeding strategies in developing hard endosperm *opaque*-2 materials. p. 37-73. *In* W.G., Pollmer and R.H., Phillips, (eds.) Improvement of Quality Traits of Maize for Grain and Silage Use, Matinus Nijhoff, London, UK.

Wallace, J.C., Lopes, M.A., Paiva, E., and Larkins,
    B.A. (1990). New methods for extraction and
    quantitation of zeins reveal a high content of
    gamma-zein in modified *opaque*-2 maize. Plant
    Physiol. 92, 191-196.
Wilson, C. (1991). Multiple zeins from maize
    endosperms characterized by reversed-phase high
    performance liquid chromatography. Plant
    Physiol. 95, 777-786.

NUTRITIONAL VALUE OF HIGH-LYSINE MAIZE IN HUMANS

Ricardo Bressani
Institute of Nutrition of Central
America and Panama
(INCAP)
Pan American Health Organization/
World Health Organization (PAHO/WHO)
Guatemala, C. A. 01011

Grain quality is a concept which is receiving attention in crop breeding programs. It is a broad concept that incorporates desirable attributes demanded by the consumer and food processor into agricultural crops. These attributes include general acceptability characteristics such as color and grain size, as well as cooking and processing quality for specific forms of consumption. To define productivity without incorporation of grain quality attributes, production by itself, is no longer a complete objective. In developing countries where human diets are made up of only a few food items, mainly of vegetable origin and in general of a relatively poor nutrient content, nutritional quality is also or should also be part of the grain quality concept. Therefore, the idea of improving the nutritional quality of basic foods such as maize, is a step forward to obtain a more complete definition of productivity. From all information available today, it can be concluded that this productivity concept has been achieved with QPM.

In this presentation, available information on the protein quality of QPM as tested in human subjects, will be reviewed. Results obtained in children will be first discussed, to be followed by results with adult human subjects.

Method of evaluation. Before reviewing information available on the protein quality of high-lysine maize, it may be useful to indicate the basic

principles of protein quality evaluation in humans.
This is mainly conducted by the nitrogen balance
method, defined in Figure 1. Nitrogen balance is the
difference between nitrogen excreted in feces and
urine from the ingested nitrogen.

When the value is positive, there occurs a
protein deposition in the body, and when negative,
protein is lost from the body. Evaluation may be
conducted at one level of protein intake, usually at
or slightly below physiological levels, or else it may
be carried out at multiple levels of intake. When
using multiple levels of protein intake, the response
between NI and NR and NA is often linear in nature up
to a certain point, which may be expressed in terms of
a simple regression equation from which a number of
values can be derived. The coefficient of regression
of NI to NA is called the nitrogen balance index,
equivalent to biological value. Most of the results
to be reviewed were obtained at a low fixed level of
intake, both in infants and children as well as in
adults.

<u>Results in children</u>. The first study herein
reported was conducted with six children who had
recovered from protein-calorie malnutrition (1). They
were from 20-60 months of age and weighed between 10.8

FIGURE 1.   SOME NITROGEN BALANCE RELATIONSHIP

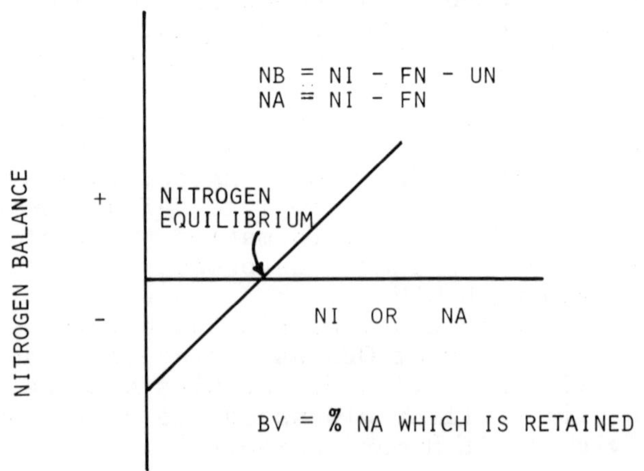

and 19.0 kg. The results are shown in Table 1. The
maize used was opaque-2, produced in Indiana by Purdue
University, and was processed by the lime-cooking
procedure to yield, after dehydration, a flour which
had a protein quality value of 96% of the casein
value, as tested in growing rats (2). Children were
fed the cooked maize as a gruel, and provided the only
dietary protein source. Calories were adjusted to 100
kcal/kg/day with vegetable fat and sugar, and a
complete vitamin and mineral mixture was given daily.
The values in the Table represent the average of three
3-day balance periods. At the higher level of protein
intake, apparent protein digestibility was lower than
that of milk protein, but NB was similar to milk. In
a second nitrogen balance study, protein intake was
decreased and, in this case, findings confirmed in
general, the results found in the first study,
although the difference to milk was greater. A third
study (1) was then undertaken with a total of 12
children, assigning 4 to each level of protein intake,
as shown in Table 2. The children ages varied from 25
to 42 months, and the study was conducted as already
explained. As protein intake from opaque-2 cooked
maize decreased, there was a slight decrease in
nitrogen balance as it usually occurs when protein
intake decreases. With these figures showing a linear
response, the nitrogen balance index was calculated
through a linear regression equation relating NA to NR
(Table 3). The coefficient of regression or NBI for
opaque-2 maize was 0.72, as compared to 0.80 for milk
protein. Using data from other human studies (3), the

Table 1. Nitrogen balance of children fed opaque-2
          maize*

Food	Protein level of intake g/kg/day	Nitrogen absorbed % intake	Nitrogen retained % intake
Milk	1.73	81.2	24.5
O-2	1.84	75.6	28.1
Milk	1.69	84.5	28.4
Milk	1.17	83.4	36.4
O-2	1.49	71.4	26.0
Milk	1.19	82.1	25.3

* As lime-treated maize. Bressani et al., 1969.

Table 2. Nitrogen balance of preschool children fed
          opaque-2 maize*

Protein level of intake g/kg/day**	Nitrogen absorbed % intake	Nitrogen retained % intake
1.87	76.3	29.0
1.49	71.4	26.0
1.21	71.5	24.3

*   Lime-treated maize.
** Four children at each level, age of 25, 42, 30 m.,
    respectively.
    Bressani et al., 1972.

Table 3. Nitrogen balance index of common and opaque-
          2 maize in children.

Food	NBI*	Protein quality % milk	Apparent % protein digestibility
Common maize	0.31	39	75.0
Opaque-2 maize	0.72	90	73.5
Milk	0.80	100	82.3

*   NB = A + B (NA); B = NBI.
    Bressani et al., 1969.

NBI of normal lime-treated maize flour was found to be
0.31. Thus, normal maize has a protein quality 37% of
the value of milk, while opaque-2 gave a value of 90%
of the value of milk protein. Common maize protein had
an apparent protein digestibility of 75.0 and opaque-2
of 73.5%, both below the value of milk. Using the
regression equations of NB to NI, calculations were
made to estimate the amount of maize of each type
needed to obtain nitrogen equilibrium, that is when NB
is equal to zero. As the data in Table 4 reveal,
calculation indicated that 23.6 g of normal maize per
kg body weight were required for nitrogen equilibrium,
while only 8.2 g were needed from opaque-2 maize. It
must be indicated that calorie intake was made
adequate by adding fat and sugar up to 100
kcal/kg/day. On the basis of 15 kg body weight for

Table 4. Maize intake for nitrogen equlibrium -
          children.

Maize	From NB studies g/kg/day	From NB studies g/day	Surveys g/day
Common*	23.6	354***	135-142
O$_2$**	8.2	124***	---

the children, nitrogen equilibrium was obtained with
354 g of normal maize, while only 124 g were required
from opaque-2 maize. Food consumption surveys
undertaken in Guatemala show maize intakes for
children 3 to 5 years of age of 135 to 142 g/child/day
(4), which from normal maize will not be enough to
obtain positive nitrogen balance. These maize intake
values are similar to the value obtained for opaque-2
maize, from the nitrogen balance study. It would be
very difficult for a small preschool child to consume
354 g of dry maize flour per day. Comparative
nitrogen retention values measured in children fed
milk, opaque-2 maize and normal maize, with and
without lysine and tryptophan supplementation, are
depicted in Figure 2 at different levels of protein
intake (3). Results show that 3 grams of protein from
normal maize supplemented with lysine and tryptophan
gave comparable NR values as opaque-2 maize fed at 1.8
g protein/kg/day. Normal maize fed at 2.0 and 1.5 g
protein/kg/day gave negative nitrogen balance, while
1.5 g protein/kg/day of opaque-2 maize induced a high
positive nitrogen balance, confirming the higher
quality of high-lysine maize.
    Other results obtained at INCAP are described in
Table 5 where the NB of other protein sources is also
presented (5). At similar levels of nitrogen intake,
milk protein gave the highest retention values,
followed by opaque-2 maize and then, significantly
less by common maize, and common maize with beans
giving similar results. It is of interest to point
out that the normal maize-bean mixture did not result
in an improved NB over that obtained from feeding
normal maize alone, even though it is known that beans
are a good supplementary protein source for maize. A
problem is the lower digestibility for maize/beans of
72% as compared to 75% from normal maize. Furthermore,

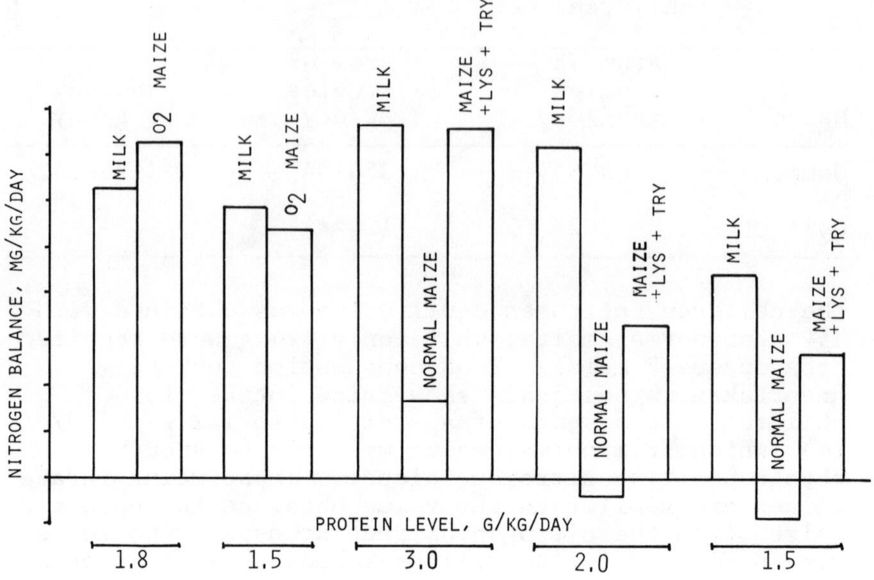

BRESSANI, 1969

Table 5. Nitrogen balance of preschool children fed diverse sources of protein*

Protein source	Nitrogen intake mg/kg/d	Nitrogen absorbed % intake	Nitrogen retained % intake
Milk	195* (173-210)	80 (61-87)	38 (22-50)
Common maize	192 (183-198)	75 (66-88)	16 (5-30)
O$_2$ maize	193 (175-200)	72 (65-76)	24 (4-34)
O$_2$ maize maize (87%):beans (13%)	207 (194-226)	72 (62-28)	17 (9-26)

* Average and range. Viteri, Martinez, Bressani, 1972.

13% beans may not provide enough lysine to show an
improvement in the quality of the mixture.

Colombia studies

Table 6 summarizes results reported by other
workers from Colombia (6). The biological value of
the opaque-2 maize and of the crystalline opaque-2
were slightly below the BV of casein, and about 1.6
times greater than the BV of normal maize, confirming
results previously shown. The same group of Colombian
investigators fed opaque-2 maize to severely
malnourished children and measured their recovery. In
these studies, with seven malnourished children,
opaque-2 maize was the only protein source. Results
from one subject are illustrated in Figure 3. In this
Figure, the stepwise line represents cumulative NB,
while the solid line represents the increase in
weight. There is a high correlation between weight
increase and cumulative NB and all 7 children behaved
alike. Recovery of the children as measured by NB,
weight increase and other parameters, was similar to
recovery of children fed with proteins of animal
origin.

Peru studies

Similar results as those already described were
reported from nitrogen balance studies conducted in
Peru (7, 8), as shown in Table 7. In this case, eight
malnourished male children, 10-25 months of age,
participated in the study. Protein intake from maize
and casein was equal to 6.4% of total calories, and
energy intake was adjusted to 100-124 kcal/kg/day.

Table 6. True protein digestibility and biological
          value of various proteins in children.

Protein	True Protein digestibility %	Biological value %
Casein	98	77
H-208 Opaque	91	76
H-208 Cryst.	87	75
H-208 Normal	78	47

Pradilla et al., 1973

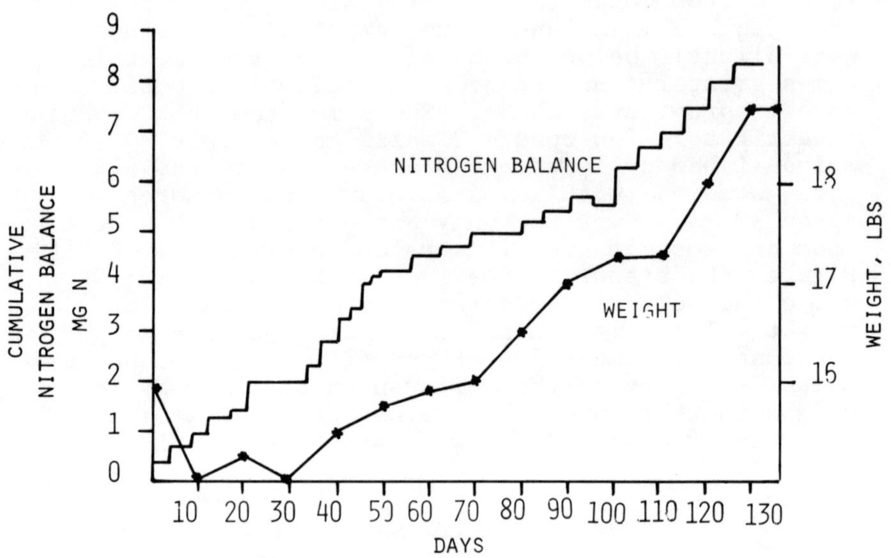

PRADILLA ET AL., 1969

Both endosperm proteins and whole kernel proteins were
tested. Besides opaque-2 maize, a sugary-2-opaque-2
maize known to contain high levels of lysine and
tryptophan was also tested. The values for endosperm
show both maize samples with the opaque- 2 gene to
give higher nitrogen retention values than the
endosperm from normal maize. This had a relative
value of 40.5% of casein, while the opaque-2 gene
endosperm had 62 and 67% the value of casein,
respectively. Higher retention values were obtained
when the whole kernel was fed, as compared to the
figures from endosperm. The samples with the opaque-2
gene superior in nutritional quality than normal

Table 7. Nitrogen balance of children fed maize
          endosperm and whole kernel flours.

Food Fed	Nitrogen balance	
	Endosperm absorbed % intake	Endosperm retained % intake
Normal	64.1 ± 11.4	15.1 ± 8.9
Opaque-2	69.6 ± 6.3	22.8 ± 5.5
Su2O2	66.7 ± 5.1	24.8 ± 8.0
Casein	81.8 ± 5.2	37.0 ± 14.2
	Nitrogen balance	
	Whole kernel absorbed % intake	Whole kernel retained % intake
Normal	73.1 ± 1.9	26.8 ± 4.6
Opaque-2	70.6 ± 5.3	30.1 ± 8.3
Su2O2	70.4 ± 4.0	31.6 ± 9.7
Casein	83.5 ± 2.5	39.6 ± 9.1*

*   Significantly different from all other values
    (P=0.05)
    Graham et al., 1980.

maize, although the differences were smaller when
compared to results obtained when endosperm proteins
were fed. This was probably due to the contribution
to protein quality provided by the germ protein, rich
in lysine and tryptophan, effect which was more
striking for normal maize. Endosperm nitrogen is rich
in zein, the protein of maize, highly deficient in
lysine and tryptophan and which contains high levels
of leucine. The lysine contribution of the germ is
more clearly seen in the results shown in Table 8. For
these calculations the digestibility figure from the
grain fed was applied to the lysine content, assuming
that lysine has the same digestibility as that for
total protein. The results indicate an increase in NR
as lysine absorbed increased, for both endosperm and
whole kernel. NR values for the whole grain are
higher, possibly because the germ also contributed to
a better essential amino acid pattern.

Table 8. Amount of lysine ingested and absorbed by
          infants consuming maize flour, related to
          apparent nitrogen retention.

Maize flour	Lysine intake mg/kg/d	Lysine absorbed mg/kg/d	Nitrogen retention mg/kg/d
Endosperm			
Normal	41 ± 5	26 ± 6	35 ± 47
O2	63 ± 7	43 ± 3	68 ± 23
Su2O2	74 ± 9	49 ± 6	75 ± 31
Casein		(100 +)	(112 ± 52)
Whole kernel			
Normal	54 ± 5	39 ± 4	81 ± 15
O2	77 ± 7	54 ± 8	93 ± 31
Su2O2	81 ± 7	60 ± 8	97 ± 35
Casein		(100 +)	(121 ± 33)

Graham et al, 1980

Studies on Nutricta

More recently, the same group of researchers
from Peru informed on nitrogen balance results using a
QPM variety called Nutricta (9). Their findings are
given in Table 9. The study was conducted as
indicated before, that is, with the participation of
six children; 6.9 to 18.5 months of age, weighing
between 4.3 to 7.7 kg. QPM gave a retention value
equivalent to 78% of the casein value, while normal
maize gave 54% of the value of casein. This QPM
variety was also tested during a 90- day growth study
conducted with malnourished children in Peru (10) with
the results illustrated in Table 10. In this study,
10 recovering malnourished children 13-29 months of
age and weighing 6.1 to 9.9 kg participated. In
general, the measures of growth and other parameters
evaluated such as kcal/g of grain, height/age change,
weight/age change, and fat-fold change were similar to
those for milk. If the values are slightly higher,
this was possibly due to a higher nitrogen intake from
QPM as compared to milk. Therefore, results confirm,
in a different way, the high quality of the protein of
QPM, very close to that of milk protein.

Table 9. Digestibility and nitrogen balance of
          infants fed common maize, QPM and casein.*

Food	N intake mg/day	N absorb. % intake	N retained % intake
Casein	2181 ± 292	82 ± 4	41 ± 9
QPM	2273 ± 295	70 ± 5	32 ± 4
Common maize	2256 ± 299	69 ± 7	22 ± 10

***From Graham et al. (1989) by permission of <u>Pediatrics</u>.**

Table 10. Effect of QPM consumed by recovering
           malnourished infants for a 90-day period.

	QPM	Milk
<u>Intake</u>		
Kcal	110 ± 15	106 ± 12
Mg N	421 ± 65	338 ± 38
<u>Measures of growth</u>		
Cm/30 d	1.23 ± 0.24	1.33 ± 0.26
g/kg/d	2.63 ± 0.62	2.6 ± 0.8
Kcal/g of gain	43.0 ± 9.2	40.3 ± 9.6
Height-Age (Mo) (Change)	5.1 ± 0.7	3.3 ± 1.2
Weight-Age (Mo) (Change)	7.5 ± 2.3	5.4 ± 1.6
Fat folds (MM) (Change)	4.8 ± 3.4	6.1 ± 3.6
Serum albumin, g/dl(final)	3.77 ± 0.31	4.00 ± 0.23

From Graham et al. (1990) by permission of <u>Pediatrics</u>.

<u>Studies in India</u>

A growth study conducted in India in 1975 is
shown in Figure 4 (11). Groups of children 18-30
months of age were assigned to different feeding
treatments after passing a physical and clinical
examination. All children were at least 60% of their
normal weight for their age. The experimental groups
included a control, a group fed milk, one fed normal
maize, and another one fed opaque-2 maize. All diets
were supplemented with calories, vitamins and
minerals, and provided 10% protein and around 405
calories. The test lasted 183 days with 25 children
in the control group, 42 on milk, 35 on normal maize,
and 32 on opaque-2 maize. The response demonstrates
weight gains from opaque-2 maize to be similar to that
of milk, and superior to the weight gain obtained from

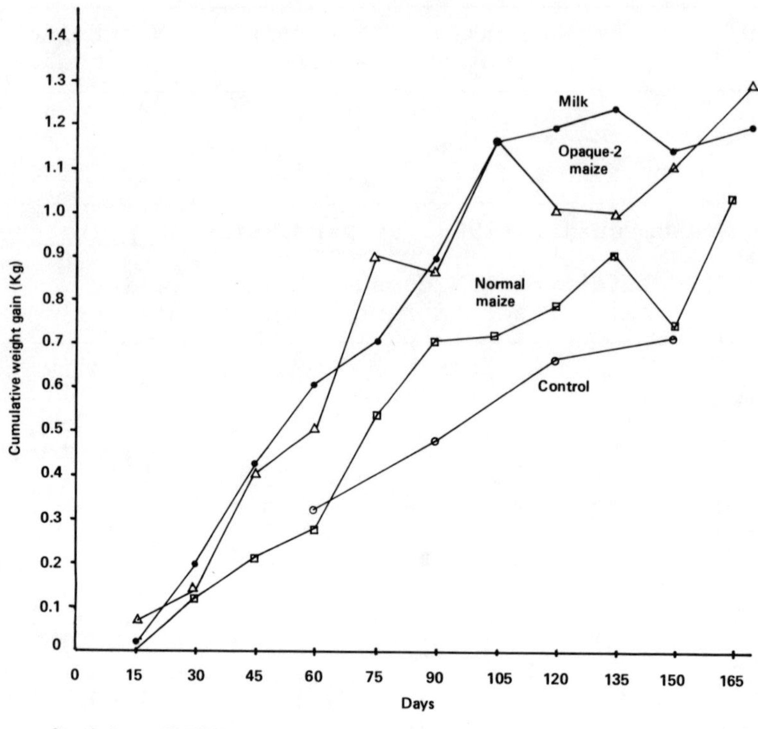

the regular diet consumed by small children, as well
as from the diet containing normal maize.

Studies with adult human subjects. Various
studies on the protein quality evaluation of opaque-2
maize in adult human subjects have been reported.
Table 11 summarizes the results of one of the first
studies, wherein four levels of opaque-2 maize were
fed as the main protein source to six subjects for
nine days at each level (12). For calculation of NB,
the N in fecal and urine samples for the last six days
was used. Total nitrogen intake was 6 g/day in all
periods. The results in the Table show that as levels
of maize decreased, nitrogen balance dropped from a
positive value of 0.29 to a negative value of minus
0.34 g. The relationship between maize intake and

216

Table 11. Nitrogen balance of adult human subjects
           fed opaque-2 maize.

Amount of maize fed g	Nitrogen balance g
330	0.29
250	0.07
200	-0.09
150	-0.34

Clark et al., 1967.

nitrogen balance was linear and this permitted the
calculation of a simple regression equation, depicted
in Table 12. From the linear equation it was
estimated that 230 g of maize were needed for nitrogen
equilibrium. In order to have a value for normal
maize, nitrogen balance data from other workers in
adult humans fed normal degerminated maize, were used
(13). The regression equation obtained is shown in
the Table for NI to NB. From the nitrogen intake at
nitrogen equilibrium the amount of normal maize was
estimated, obtaining a figure of 547 g. It is
recognized that degerminated maize has a lower protein
quality value; however, these data serve to have a
better perception of the superior quality of QPM. In
another study -the results of which are given in Table
13- five human adults were fed 4 g nitrogen per day
from three sources: a normal maize, whole opaque-2
maize and degerminated opaque-2 maize (14). All
nitrogen balances were negative, but significantly
less for the opaque-2 maize, with or without the germ.
An additional study was also conducted wherein rather
than maintaining nitrogen intake constant, maize flour
intake was kept equal at 222 g/day, which provided
different levels of nitrogen intake. As in the
previous study, nitrogen balance values from the
samples with the opaque- 2 gene were significantly
less negative. Thus, on the basis of equal N intake
and on the basis of equal intake of maize flour, the
two opaque-2 lines were shown to promote superior
nitrogen retention in comparison with normal high-
protein maize. Other results from several authors
(15, 16, 17) are summarized in Table 14. They show
true protein digestibility of opaque-2 maize to be
92%, with BV of 80, compared to 96 and 86% for egg,
repectively. The Table also cites BV in normal maize

from other laboratories, showing figures which vary
from 40 to 57%. Table 15 summarizes the important
values of protein quality evaluation in children and
adults. As observed, all values are remarkably close,
and clearly demonstrate the superiority of high-lysine
maize over normal maize in studies of its protein
nutritive value in human subjects.

Table 12. Amount of maize required by human adults
           for nitrogen equilibrium.

	Opaque-2 maize g/day	Normal maize*** endosperm g/day
Regression equation NB = 4.10 (g maize) -0.943 for NB = 0	230	-
Regression equation NB = -2.625 + 0.381 (NI) for NB = 0	-	547

*   Clark et al., 1967.
** Kies et al., (8% protein in maize endosperm).

Table 13. Comparative protein value of three lines of
           maize for adult humans.

Maize type	Nitrogen at equal nitrogen intake (4g/day)	Balance at equal intake of maize (222 g/day)*
Normal single cross hybrid (High protein)	-1.58a	-1.91a
Whole opaque-2 maize	-1.00b	-1.00b
Degerm opaque-2 maize	-0.69b	-1.24b

*   This intake provided 4.45 g N/day, 4.80 g N/day
    and 4.15 g N/day, respectively.
Kies and Fox, 1972.

Table 14. Biological value of QPM and of normal in
          adult humans.

Food	True Protein Digest %	Biol Value %	Reference
O$_2$	92	80	Young et al., 1971
Egg	96	86	
Normal	-	40	Trunswell & Block,1962
Normal	-	57	
Normal endopserm	-	46	Kies et al., 1965

Table 15. Protein quality values of high-lysine and
          normal maize.

	Hi-Lys Maize*	Normal Maize*	Reference Protein*
App. Protein Digest., %	73.5	75.0	82.3 (Milk)
True Protein Digest., %	89.0	78.0	98.0 (Casein)
Biological value	75.5	47.0	77.0 (Casein)
Amount needed for N equil. g/day	124	547	-

	Hi-Lys Maize**	Normal Maize**	Reference Protein**
App. Protein Digest., %	76.5	75-80	-
True Protein Digest., %	92	82-91	96 (egg)
Biological value	80	40-57	86 (egg)
Amount needed for N equil. g/day	230	547	-

* In children; ** In adults.

        <u>Other nutritional effects observed from feeding
high-lysine maize.</u> Although the most important
nutritional characteristics of high-lysine maize is
its protein quality, other nutritional benefits have
also been observed. These are listed in Table 16 and
include stimulation of food intake, thus increasing
protein and calorie intake (2), a higher niacin
availability probably due to a higher tryptophan and
lower leucine content (2), a higher calcium
availability when fed in the form of lime-treated
maize (18), a higher carotene bioutilization in yellow
maize (19), and a high carbohydrate utilization (10).
Furthermore, it is a better supplement than normal
maize to bean diets.

Table 16. Effects of improved lysine and content of
           maize other than higher protein quality.

---

1.   Stimulation of food intake
2.   Higher niacin availability
3.   Higher calcium bioutilization
4.   Higher carotenoid bioutilization
5.   Higher CHO bioutilization
6.   Better supplement to bean diets.

        <u>Acceptability and nutritional impact</u>. In 1976 a
field study was designed in Guatemala for the purpose
of evaluating the impact of changing normal maize for
protein quality maize in the field (20). To conduct
the study, nine coffee plantations were selected in
the region of Patulul. It is not possible to describe
all the details in this presentation, but the study
included experiences in high-lysine maize production,
storage, nutritional evaluation when raw and
processed, acceptability by families, and a 17- month
period of observation on the impact high-quality maize
had in children. Table 17 presents data on processing
acceptability by households given soft and hard
endosperm-high- lysine maize and normal maize. The
parameters measured, shown in the Table, clearly
demonstrate no differences between the two QPM
varieties and normal maize. Acceptability by the
households is summarized in Table 18. Due to the
large amounts of data, only the highest evaluation
value for each acceptability characteristic is shown.
In general, these results indicate that the members of
the households found the QPM tortillas similar to the

Table 17.   Processing Acceptability of three varieties of maize.   The Patulul field study.

	Soft 02	Hard 02	Normal
# families	19	20	20
Kg given	1.47± 0.24	1.54± 0.22	1.62± 0.34
Lime added g/kg	14.4 ±17.8	58.5 ±37.7	54.0 ±29.5
Water added lt/kg	1.7 ± 0.5	1.8 ± 0.5	1.8 ±0.6
Cooking time, min.	54.5 ±27.0	53.3 ±24.1	51.0 ±27.7
Kg cooked maize/kg dry maize	2.10± 0.27	2.15± 0.36	1.91± 0.29
# tortilla/ 1 kg	40.9 ±11.9	41.0 ±12.4	38.0 ±11.0

Valverde et al., 1983.

Table 18.   Acceptability of families of 3 varieties of maize.   The Patulul field study (%)

	Soft 02	Hard 02	Normal
Appearance of Maize			
Better than usual	78.9	80	75
Yield of tortillas			
More	73.7	55	55
Quality of tortillas			
Different but tasty	89.5	80	80
Acceptability by household:			
- Lady	100	100	100
- Head	100	100	100
Reasons:			
(Soft)			
- Lady	68.4	60	80
- Head	73.7	45	55

Valverde et al., 1983.

normal maize tortillas, and were appreciated because they were soft in texture and flexible. Furthermore, the number of tortillas produced per kg of raw maize was comparable between maize samples. The conclusion of the short-time study is given in Table 19, presented as such because of the difficulty to summarize the large amount of data accumulated and analyzed. The results obtained suggest that

Table 19.   Nutritional impact of high protein maize.
            The Patulul project.

"The results, describing the impact of Opaque-2 Maize
in children's nutritional status, suggest that the
replacement of normal by Opaque-2 maize varieties
results in improvements on children's growth."

Opaque-2 maize should be a component of other
interventions modifying living conditions as medical
care, improvement of water supplies, increases in
income, nutrition education"

Valverde et al., 1983.

replacement of normal by opaque-2 maize results in an
improvement of children's growth. Nevertheless, it
should be clear that opaque-2 maize must be a
component of other interventions modifying living
conditions, providing medical care and nutrition
education, and introducing water supplies among a
number of other components of the intervention.

CONCLUSIONS

The data -which demonstrate the superiority of
QPM over normal maize in human nutrition- are
overwhelming. Likewise, the results of many studies
have shown that food products based on maize have
equal functional properties when QPM is used, but with
improved nutritive value. These food products include
infant cereal foods highly important in the weaning
process to reduce the risk of undernutrition in
children. In all children studies reviewed in this
document, QPM protein was the only protein source fed
to children recovering from malnutrition, and this
resulted in an increased nitrogen balance and
improvement in their health condition. This fact
strongly suggests that QPM can be a practical solution
to home-made weaning foods. Unfortunately, aggressive
programs to introduce QPM in commercial production
have been very limited to date.

LITERATURE CITED

1.   Bressani, R., J. Alvarado, F. Viteri.
      Evaluación en niños de la calidad de la proteína

del maíz opaco-2. Arch. Latinoamer. Nutr. 19:129-140, 1969.

2. Bressani, R., L. G. Elías, R. A. Gómez-Brenes. Protein quality of opaque-2 corn. Evaluation in rats. J. Nutr. 97:173-180, 1969.

3. Bressani, R. Amino acid supplementation of cereal grain flours tested in children. p. 184-204. In: Amino Acid Fortification of Protein Foods. Eds. N. S. Scrimshaw & A. M. Altschul, 1971. The MIT Press, Cambridge, MA. USA.

4. Flores, M., Z. Flores, M. Y. Lara. Food intake of guatemalan indian children, ages 1-5. J. Amer. Diet. Assoc. 48:480-487, 1966.

5. Viteri, F., C. Martínez, R. Bressani. Evaluation of the protein quality maize, opaque-2 maize and common maize supplemented with amin acids and other sources of protein. p. 191-204. In: Nutritional Improvement of Maize. Eds. R. Bressani, J. E. Braham & M. Behar. INCAP. Publication L- 4. October 1972. Guatemala.

6. Pradilla, A., F. Linares, C. A. Francis, L. Fajardo. El maíz de alta lisina en nutrición humana. p. 41-48. In: Simposio Sobre Desarrollo y Utilización de Maíces de Alto Valor Nutritivo. Colegio de Postgraduados, ENA, Chapingo, Mexico, 1973.

7. Graham, G. G., D. V. Glover, C. López de Romaña, E. Morales, W. C. MacLean Jr. Nutritional value of normal, opaque-2 and sugary-2 maize hybrids for infants and children. 1. Digestibility and utilization. J. Nutr. 110:1061-1069, 1980.

8. Graham, G. G., R. P. Placko, W. C. MacLean Jr. Nutritional value of normal, opaque-2 and sugary- 2 maize hybrids for infants and children. II. Plasma-free amino acids. J. Nutr. 110:1070-1075, 1980.

9. Graham, G. G., J. Lembcke, E. Lancho, E. Morales. Quality protein maize: digestibility and utilization by recovering malnourished infants. Pediatrics 83:416-421, 1989.

10. Graham, G. G., J. Lembcke, E. Morales. Quality protein maize as the sole source of dietary protein and fat for rapidly growing young children. Pediatrics 85:85-91, 1990.

11. Singh, J. Studies on assessing the nutritive value of opaque-2 maize. Final report of the Project Indian Agricultural Research Institute, New Delhi, India, 1977.

12. Clark, H. E., P. E. Allen, S. M. Meyers, S. E. Tuckett, Y. Yamamura. Nitrogen balance of adults consuming opaque-2 maize protein. Amer. J. Clin. Nutr. 20:825-833, 1967.

13. Kies, C.; E. Williams, H. M. Fox. Determination of first limiting nitrogenous factor in corn proteins for nitrogen retention in human adults. J. Nutr. 86:350-356, 1965.

14. Kies, C., H. M. Fox. Protein nutritional value of opaque-2 corn grain for human adults. J. Nutr. 102:757-765, 1972.

15. Young, V. R., I. Ozalp, B. V. Chalakos, N. S. Scrimshaw. Protein value of colombian opaque-2 corn for young adult men. J. Nutr. 101:1475-1481, 1971.

16. Clark, H. E., D. V. Glover, J. L. Betz, L. B. Bailey. Nitrogen retention of young men who consumed isonitrogenous diets containing normal, opaque-2 or sugary-2-opaque-2 corn. J. Nutr. 107:404-411, 1977.

17. Truswell, A. S., J. F. Brock. The nutritive value of maize protein for man. Amer. J. Clin. Nutr. 10:142-146, 1962.

18. Braham, J. E., R. Bressani. Utilización del calcio del maíz tratado con cal. Nutr. Bromatol. Toxicol. 5:14-19, 1966.

19. De Bosque, C., E. J. Castellanos, R. Bressani. Biodisponibilidad y digestibilidad de carotenoides de maíz común y maíz suplementado con lisina y triptofano en ratas. INCAP Annual Report 1988.

20. Valverde, V., H. L. Delgado, J. M. Belizán, R. Martorell, V. Mejía-Pivaral, R. Bressani, L. G. Elías, M. Molina, R. E. Klein. The Patulul Project: production, storage, acceptance and nutritional impact of opaque-2 corns in Guatemala. 1983. INCAP, Guatemala.

QPM AS A SWINE FEED

Darrell A. Knabe, J. S. Sullivan and K. G. Burgoon
Animal Science Department

A. J. Bockholt
Soil and Crop Sciences

Texas A&M University
College Station, Texas 77840

INTRODUCTION

Cereals are the basis of swine diets due to the
pig's ability to efficiently use starch as an energy
source. Cereals are low to moderate in protein
content, and because the quality of their protein is
poor, high-protein feedstuffs must be added to cereals
if adequate pig performance is to be obtained.
Soybean meal is the most common high-protein feedstuff
fed to swine in the USA, due to its cost, abundance
and its amino acid pattern which complements the amino
acid pattern of cereals. Specifically, soybean meal
is high in lysine, tryptophan and threonine, the
limiting amino acids in most cereals. Unfortunately,
soybean meal is two to three times more expensive than
cereals. A cereal that is an excellent source of
energy and contains a large amount of quality protein
would be an economic asset to swine producers.
Quality Protein Maize offers such a possibility.
Its amino acid profile is superior to that of normal
corn, and its hard endosperm offers improved agronomic
and storage characteristics over conventional opaque-2
corn. The utility of opaque-2 corn in swine diets has
been well demonstrated, but the nutritional value of
QPM for swine has not been as thoroughly investigated.
This paper will focus on research (Sullivan et al.,
1989; Burgoon et al., 1992) conducted at Texas A&M
evaluating QPM in swine diets.

Corns

Seven experiments were conducted over a two-year period to evaluate the nutritional value of QPM for swine. Experiments 1 to 3 were conducted the first year and evaluated QPM, normal (Mo17 x B73), and food (Asgrow 404) corns grown under the same agronomic conditions. Experiments 4 to 7 were conducted the second year. The QPM and normal corns used in these experiments were the same as those used the first year, but were grown one season later. Conventional opaque-2 corn, and two high-protein corns (HP1 and HP2) were obtained from private sources.

Growth Trials

The four growth trials (Exp. 1, 2, 4 and 5) followed a randomized complete block design, with blocks formed on the basis of initial weight (starter pigs) or initial weight and sex (grower and finisher pigs). Pigs received feed and water ad libitum throughout the experiments.

Dietary treatments were chosen to determine if the higher lysine and tryptophan contents of QPM can be used to lower the level of dietary soybean meal. For example, in Exp. 1, the QPM diet contained 19.66% soybean meal and .96% dietary lysine; a normal corn diet also contained .96% lysine, but contained 23.97% soybean meal. A normal corn diet containing the level of soybean meal in the QPM diet was also included to determine if the lysine level chosen for the QPM diet was above the pig's requirement.

Experiments 1 (starter pigs) and 2 (grower pigs) had the same arrangement of treatments. In addition to evaluating QPM, the trials also compared normal and food corn. Experiments 4 (starter) and 5 (finisher pigs) compared QPM and normal corn at two levels of soybean meal supplementation. The low level of soybean meal supplementation combined with QPM, and the high level of soybean meal supplementation combined with normal corn produced the same dietary lysine level. Experiment 5 also evaluated the feasibility of feeding QPM with supplemental crystalline lysine and tryptophan to finishing pigs. This diet contained the least amount of soybean meal.

Digestion Trials

The initial digestion trial (Exp. 3) determined total tract nutrient digestibilities and N balance data when normal corn, food corn and QPM-based diets contained 3.25% casein. Experiments 6 and 7 used pigs fitted with ileal T-cannula and determined apparent ileal digestibilities of amino acids of the corns. The corns were the only source of protein in these diets. Experiments 3 and 7 were conducted as Latin squares; Exp. 6 followed a cross-over design.

RESULTS AND DISCUSSION

Characterization of the Corns

The QPM had the highest density, and volumetric weight was greater for QPM than normal or conventional opaque-2 corns verifying that the QPM used in these studies had a hard endosperm (Table 1). As expected, total protein, lysine and tryptophan contents were higher for QPM than normal corn, and QPM and opaque-2 corn had more similar contents. The high protein corns contained the greatest amount of protein, but their lysine and tryptophan contents did not differ greatly from those of normal corn. The increased protein content resulted primarily from increased glutamic acid and leucine content. On average, QPM contained about 40% more lysine and tryptophan than normal corn.

Experiments 1 to 3

Results of these experiments suggest that QPM has higher nutritional value than food and feed corn due to increased protein quality. In the growth trials (Table 2), pigs fed the QPM diet utilized feed more (P < .10 or P < .01) efficiently (Exp. 1 and 2) and grew faster (P < .01, Exp. 2) than pigs fed food and feed corn diets containing the same level of soybean meal supplementation. Nitrogen retention expressed as a percentage of N intake or as a percentage of N absorbed was also highest (P < .05) for QPM (Table 3). The amino acid or amino acids responsible for this higher nutritional value

TABLE 1. CHARACTERIZATION OF THE CORNS[a]

Item	Exp. 1 to 3			Exp. 4 to 7				
	QPM	Food	Normal	QPM	Normal	0-2[b]	HP1	HP2
Volumetric weight, kg/l	.77	.79	.75	.79	.75	.71	.79	.77
Density, g/cm^3	1.34	1.33	1.28	1.31	1.25	1.18	1.30	1.29
Moisture, %	10.9	12.0	10.6	11.1	12.7	9.4	11.9	8.7
CP (N x 6.25), %	10.2	8.8	7.6	9.8	9.1	8.9	11.5	11.1
Ether extract, %	4.8	5.0	3.8	4.4	3.2	4.5	4.1	3.9
Ash, %	2.40	1.72	1.28	1.5	1.5	1.2	1.4	1.4
GE, kcal/g	4.04	4.05	4.01	3.99	3.99	4.08	4.11	4.15
Arginine	.71	.46	.43	.66	.51	.59	.53	.56
Histidine	.31	.24	.21	.37	.28	.31	.34	.34
Isoleucine	.28	.27	.24	.32	.34	.30	.42	.41
Leucine	.66	.83	.71	.91	1.14	.80	1.61	1.51
Lysine	.45	.31	.29	.40	.31	.40	.31	.32
Methionine	.19	.18	.17	.17	.17	.14	.29	.21
Phenylalanine	.34	.35	.30	.41	.47	.39	.61	.59
Threonine	.34	.29	.26	.36	.35	.32	.40	.39
Tryptophan	.083	.055	.054	.074	.054	.065	.055	.065
Valine	.48	.40	.36	.52	.46	.48	.57	.58

[a]Values are on an as fed basis.
[b]Conventional opaque-2 corn.

TABLE 2. PERFORMANCE OF STARTER AND GROWER PIGS FED DIETS CONTAINING QPM, FOOD OR NORMAL CORN (EXP. 1 AND 2)

| | | Low SBM | | High SBM | | Contrast | | | | |
Item	QPM (1)	Food (2)	Normal (3)	Food (4)	Normal (5)	1 vs 2+3	1 vs 4+5	2+4 vs 3+5	2+3 vs 4+5	CV
Exp. 1[a]										
Dietary lysine,%	.96	.85	.83	.97	.96					
Dietary soybean meal,%	19.66	19.66	19.66	23.97	23.97					
Avg daily intake,kg	.57	.61	.55	.61	.59					9.7
Avg daily gain,kg	.30	.30	.27	.33	.32		*	+	**	7.0
Feed/gain	1.92	2.04	2.04	1.85	1.85	+			**	5.8
Exp. 2[b]										
Dietary lysine,%	.70	.58	.56	.72	.70					
Dietary soybean meal,%	9.95	9.95	9.95	14.83	14.83					
Avg daily intake,kg	1.97	1.92	1.86	1.93	1.92					7.5
Avg daily gain,kg	.75	.66	.63	.77	.75	**			**	10.5
Feed/gain	2.63	2.94	2.94	2.50	2.56	**	*		**	4.8

[a]Values are means of four pens with six pigs each. The pigs initially averaged 5.8 kg, and the trial lasted 28 d.

[b]Values are means of eight pens with two pigs each. The pigs initially averaged 22.8 kg, and the trial lasted 35 d.

+ P < .10. * P < .05. **P < .01.

TABLE 3. APPARENT DIGESTIBILITIES OF DIET COMPONENTS, ENERGY VALUES OF DIETS AND NITROGEN BALANCE OF PIGS FED QPM, FOOD OR NORMAL CORN (EXP.3)[a]

Item	QPM	Food	Normal	CV
Dry matter digestibility, %	89[d]	90[c]	89[d]	.8
Gross energy digestibility, %	90	90	89	1.3
Metabolizable energy, kcal/g[b]	3.87[c]	3.88[c]	3.80[d]	.9
N digestibility, %	89[c]	87[c]	85[d]	1.6
N balance				
Intake, g/d	33.1	28.5	26.0	
Retained, g/d	19.6[c]	12.0[d]	11.2[d]	6.1
Retained, % of intake	59.1[c]	42.1[d]	43.4[d]	4.2
Retained, % of absorbed	66.6[c]	48.4[d]	51.3[d]	4.4

[a]Diets contained 3.25% casein. Digestibilities and energy contents were calculated from total feces and urine collection data.

[b]Dry matter basis.

[cd]Values in the same row not sharing a common superscript letter differ (P <. 05).

cannot be determined from these experiments, but the higher lysine and tryptophan content of QPM are probably responsible. The higher feeding value for opaque-2 corn than normal corn has been attributed to its higher lysine and tryptophan content (Cromwell et al., 1967; Klein et al., 1971). Opaque-2 corn has also been shown to support higher N retention than normal corn (Cromwell et al., 1969).

In Exp. 2, growth rate of pigs fed QPM was equivalent to that of pigs fed isolysinic feed and food corn diets, but efficiency of feed utilization was slightly less. In Exp. 1, growth rate was lower for pigs fed the QPM diet than for pigs fed isolysinic food and feed corn diets. These results suggest that QPM, although superior to conventional corn, cannot be fed in soybean meal supplemented starter and grower diets formulated to .96 and .70% lysine, respectively, and obtain performance equivalent to that obtained on normal corn-soybean meal diets containing the same level of lysine.

The cause of this lowered performance is not evident. The similar energy digestibilities of the diets determined in Exp. 3, and the high digestible energy content of the QPM diet, suggest that differences in energy utilization should not have been a factor. Isoleucine and threonine contents were lower for the QPM diets than the food and feed corn diets. Isoleucine contents were adequate in both diets based on NRC (1988) requirements. Threonine was adequate in the grower diet, but slightly marginal in the starter diet. Whether these differences are responsible for the lowered pig performance on the QPM diets cannot be determined from the present study and merits further research. It should be noted that all diets were deficient in lysine, and the use of higher soybean meal supplementation, which would occur in commercial diets, would minimize any possible amino acid shortages in QPM diets. For this reason, the lysine levels of diets used in Exp. 4 and 5 more closely match those used in commercial diets.

Experiments 4 to 7.

No differences in performance were found among starter pigs fed normal corn or QPM-based diets

supplemented with two levels of soybean meal (Table
4).   The lack of improved performance in starter pigs
fed QPM may have resulted from a surplus of lysine in
the QPM diets rather than an inability of the pigs to
utilize QPM.   Increasing soybean meal in the normal
corn diet increased dietary lysine from .99 to 1.05%,
but did not affect pig performance, suggesting that
.99% lysine was at or above the requirement of pigs
fed simple corn-soybean meal diets.   Thus, a
difference in performance between pigs fed the low-
soybean meal-QPM diet (1.05% lysine) and pigs fed the
low-soybean meal-normal corn diet (.99% lysine) would
not be expected.   Prior trials (Cromwell et al., 1967;
Sihombing et al., 1969; Marroquin et al., 1973; Asche
et al., 1985) evaluating conventional opaque-2 corn,
found that starter pigs fed opaque-2 corn had
performance superior to that of pigs fed normal corn
when diets were supplemented with soybean meal to
provide suboptimal protein levels.   In Exp. 1, pigs
fed the QPM diet had reduced growth rate compared to
pigs fed isolysinic (.96%) normal corn diet.   In the
present study, pigs fed isolysinic (1.05%) normal and
QPM diets had similar performance.

Finishing pigs fed the QPM diet with the low
level of soybean meal consumed more (P < .05) feed and
had higher (P < .05) rates of gain than pigs fed the
normal corn diet with the same level of soybean meal
supplementation (Table 4).   Feed efficiency also
tended (P < .15) to improve on the QPM diet.   Rate and
efficiency of gain was equal on this QPM diet and the
normal corn diet containing the same dietary lysine
level.   Increasing soybean meal content in the normal
corn diet improved average daily gain (P < .05) and
feed efficiency (P < .01).   A similar increase in
soybean meal content in the QPM diet did not seem to
improve performance.   This differential response to
increasing soybean meal content in the QPM and normal
corn diets was responsible for the corn type X soybean
meal level interaction that approached significance (P
< .15).

Pigs fed the low-protein QPM diet supplemented
with lysine and tryptophan did not perform as well as
pigs fed the normal corn diet with the higher level of
soybean meal; daily gains tended to be reduced and
feed efficiency was less (P < .05).

These data suggest that for finishing pigs less
protein supplementation is needed in QPM-based diets

TABLE 4. PERFORMANCE OF STARTER AND FINISHER PIGS FED DIETS CONTAINING QPM OR NORMAL CORN (EXP. 4 and 5)

| | Low SBM | | High SBM | | QPM | Contrasts | | | | |
| | Normal | QPM | Normal | QPM | Lys+Trp | 1 vs | 2 vs | 3 vs | 3 vs | |
	(1)	(2)	(3)	(4)	(5)	2	3	4	5	CV
Exp. 4[a]										
Dietary lysine,%	.99	1.05	1.05	1.11						
Dietary soybean meal	26.53	26.53	28.96	28.96						
Average daily intake,kg	.47	.48	.48	.51						5.5
Average daily gain,kg	.29	.29	.30	.30						7.4
Feed/gain	1.64	1.65	1.60	1.71						5.7
Exp. 5[b]										
Dietary lysine,%	.58	.66	.66	.74	.66					
Dietary soybean meal	10.80	10.80	13.80	13.80	6.00[c]					
Average daily intake,kg	3.19	3.43	3.31	3.37	3.31	*				6.5
Average daily gain,kg	.93	1.03	1.04	1.05	.98	*		*		7.4
Feed/gain	3.44	3.31	3.19	3.22	3.36			**	*	4.6

[a]Values are means of five pens with six pigs each. The pigs initially averaged 6.8 kg and the trial lasted 21 d.

[b]Values are means for seven pens with two pigs each. The pigs initially averaged 59 kg and the trial lasted 42 d.

[c]Diet supplemental with .15% lysine HCl, 98% and .05% DL-tryptophan.

*P < .05.

**P < .01.

than in normal corn-based diets to maximize
performance. This agrees with results of Exp. 2 and
with prior research evaluating conventional opaque-2
corn (Cromwell et al., 1969; Gipp and Cline, 1972;
Asche et al., 1985) in diets of growing and finishing
pigs.

Finishing pigs fed the QPM diet supplemented
with lysine and tryptophan did not perform as well as
pigs fed the normal corn diet supplemented with the
high level of soybean meal, suggesting a deficiency of
an essential amino acid. Which amino acid may have
been limiting is speculative, since the diet was
formulated to contain 110% of the NRC (1988)
requirements for lysine, tryptophan, threonine and
isoleucine. All other essential amino acid contents
were 124% (methionine + cystine) or greater than the
NRC (1988) requirement. Wahlstrom et al. (1977)
reported that lysine was first limiting, and threonine
second limiting in opaque-2 corn for starter pigs;
additions of methionine, tryptophan and isoleucine did
not improve performance. Oestemer et al. (1970) also
found that the addition of methionine to an opaque-2
corn diet supplemented with lysine, tryptophan,
threonine and isoleucine did not improve performance
of growing pigs. Jensen et al., (1969) found that
opaque-2 corn supplemented with lysine alone supported
performance of finishing pigs that was equal to that
obtained on a 12% protein normal corn-soybean meal
diet. The inability of the low protein, amino acid
supplemented QPM diet to support maximum performance
contrasts with the reports of Jensen et al. (1969),
Gipp and Cline (1972) and Veum et al. (1973) who
reported that opaque-2 corn supplemented with amino
acids resulted in performance equal to control diets.

Apparent digestibilities of lysine and
tryptophan were higher (P < .01) for QPM than normal
corn in Exp. 6 (Table 5). Based on plasma amino acid
levels (Pick and Meade, 1970) and slope-ratio assay
with rats (Klein et al., 1972) lysine in opaque-2 corn
was more available than lysine in normal corn.
Riviera et al. (1978) also found slight but consistent
improvements in the apparent digestibility of amino
acids in opaque-2 corn compared to normal corn using
the fecal analysis method.

Quality Protein Maize and opaque-2 corn had
higher (P < .05) apparent lysine digestibilities than
the two high-protein corns. Tryptophan

TABLE 5. APPARENT DIGESTIBILITIES OF GROSS ENERGY, N AND AMINO ACIDS IN THE CORNS
(EXP. 6 AND 7)[a]

	Exp. 6				Exp. 7				
	QPM	Normal	P	CV	QPM	Opaque-2	HP1	HP2	CV
Gross energy[b]	89	89		.9	88[d]	87[e]	89[d]	89[d]	.9
N[c]	74	73		2.4	75[e]	73[f]	77[d]	75[e]	1.3
Arginine[c]	86	82	*	1.5	87[d]	82[f]	85[e]	84[e]	.7
Histidine	86	83	**	1.6	86[d]	83[e]	86[d]	85[d]	.9
Isoleucine	75	74		3.6	75[f]	73[g]	80[d]	78[e]	.8
Leucine	84	87	**	.7	80[e]	79[e]	85[d]	85[d]	1.2
Lysine	72	67	**	3.3	74[d]	74[d]	67[e]	67[e]	1.8
Methionine	79	83	*	2.5	79[d]	75[e]	82[d]	79[d]	2.4
Phenylalanine	84	85	*	.7	82[e]	79[f]	85[d]	85[d]	.9
Threonine	67	64		3.9	69[e]	66[f]	72[d]	70[e]	1.1
Tryptophan	74	63	**	5.6	74[d]	65[e]	66[e]	61[f]	3.3
Valine	79	76	*	1.9	80[de]	77[f]	81[d]	79[e]	.9

[a]Values are percentages and are means of six observations for Exp. 6 and four observations for Exp. 7. Corn was the only source of dietary protein in all diets.

[b]Determined from collection of feces.

[c]N and amino acid digestibilities determined from collection of ileal digesta.

[def]Values with different superscripts in the same row differ (P < .05).

*P < .05. **P < .01.

235

digestibilities were highest (P < .05) for QPM,
similar for opaque-2 corn and HP1 and lowest (P < .05)
for HP2. It should be noted that apparent amino acid
digestibilities can be affected by amino acid content
of the diet; i.e., as amino acid content increases,
apparent digestibility increases because the
proportion of endogenous amino acids in the excreta
becomes less. This phenomenon may partially explain
the differences in apparent amino acid digestibility
among the corns. In all instances, except methionine,
differences in apparent digestibilities between QPM
and normal corn parallel differences in amino acid
content. This effect was also found among QPM,
opaque-2 corn and the high-protein corns where amino
acids contents of the corns varied the most
(isoleucine, leucine, lysine, phenylalanine).
Overall, it appears the corns evaluated in this
research have essentially the same digestibilities.

CONCLUSION

Energy and amino acids in QPM are as digestible
as those in normal corn or conventional opaque-2 corn.
The higher lysine and tryptophan content of QPM is
effectively utilized by pigs for growth. Thus, diets
based on QPM require less protein supplementation
which should provide an economic incentive for use of
QPM in swine diets.

LITERATURE CITED

Asche, G. L., A. J. Lewis, E. R. Peo, Jr. and J. D.
    Crenshaw. 1985. The nutritional value of normal
    and high lysine corns for weanling and growing-
    finishing swine when fed at four lysine levels. J.
    Anim. Sci. 60:1412.
Burgoon, K. G., J. A. Hansen, D. A. Knabe and A. J.
    Bockholt. 1992. Nutritional value of Quality
    Protein Maize for starter and finisher swine. J.
    Anim. Sci. (in press).
Cromwell, G. L., R. A. Pickett and W. M. Beeson.
    1967. Nutritional value of opaque-2 corn for
    swine. J. Anim. Sci. 26:1325.
Cromwell, G. L., R. A. Pickett, T. R. Cline and W. M.
    Beeson. 1969. Nitrogen balance and growth studies
    of pigs fed opaque-2 and normal corn. J. Anim.
    Sci. 28:478.

Gipp, W. F. and T. R. Cline. 1972. Nutritional
    studies with opaque-2 and high protein opaque-2
    corns. J. Anim. Sci. 34:963.
Jensen, A. H., D. H. Baker, D. E. Becker and B. G.
    Harmon. 1969. Comparison of opaque-2 corn, milo
    and wheat in diets for finishing swine. J. Anim.
    Sci. 29:16.
Klein, R.G., W.M. Beeson, T.R. Cline and E.T. Mertz.
    1971. Opaque-2 and floury-2 corn studies with
    growing swine. J. Anim. Sci. 32:256.
Klein, R. G., W. M. Beeson, T. R. Cline and E. T.
    Mertz. 1972. Lysine availability of opaque-2 corn
    for rats. J. Anim. Sci. 35:551.
Marroquin, C. R., G. L. Cromwell and V. W. Hays.
    1973. Nutritive value of several varieties of
    opaque-2 corn and normal corn for growing swine.
    J. Anim. Sci. 36:253.
NRC. 1988. Nutrient Requirements of Swine (9th Ed.).
    National Academy Press, Washington, DC.
Oestemer, G. A., R. J. Meade, W. L., Stockland and L.
    E. Hanson. 1970. Methionine supplementation of
    opaque-2 corns for growing swine. J. Anim. Sci.
    31:1133.
Pick, R. I. and R. J. Meade. 1970. Nutritive value
    of high lysine corn: Deficiencies and
    availabilities of lysine and isoleucine for growing
    swine. J. Anim. Sci. 31:509.
Riviera, L. P. H., R. E. Peo Jr., D. Flowerday, T. D.
    Crenshaw, B. D. Moser and P. J. Cunningham. 1978.
    Effect of maturity and drying temperature on
    nutritional quality and amino acid availability of
    normal and opaque-2 corn for rats and swine. J.
    Anim. Sci. 46:1024.
Sihombing, T. H., G. L. Cromwell and V. W. Hays.
    1969. Nutritive value and digestibility of opaque-
    2 and normal corn for growing pigs. J. Anim. Sci.
    29:921.
Sullivan, J. S., D. A. Knabe, A. J. Bockholt and E. J.
    Gregg. 1989. Nutritional value of Quality Protein
    Maize and food corn for starter and grower pigs.
    J. Anim. Sci. 67:1285.
Veum, T. L., W. H. Pfander, C. G. Bellamy and H. B.
    Hedrick. 1973. Opaque-2 and normal corn as amino
    acid sources for barrows and gilts. J. Anim. Sci.
    36:877.
Wahlstrom, R. C., R. V. Merrill, L. J. Reiner and G.
    W. Libal. 1977. Mutant corns in young pig diets

and amino acid supplementation of opaque-2 corn.
J. Anim. Sci. 45:747.

WET MILLING PROPERTIES OF QUALITY PROTEIN MAIZE AND REGULAR CORNS

M.H. Gomez, S.O. Serna-Saldivar, J.I.
Corujo, A.J. Bockholt, and L.W. Rooney.
Cereal Quality Lab.
Dept. Soil and Crop Sciences.
Texas A&M University,College Station, TX
77843-2474

INTRODUCTION

Lysine and/or tryptophan limit the efficiency of corn protein utilization. Scientists throughout the world have worked extensively to develop cereals with improved protein quality (Mertz et al., 1964; Singh and Axtell, 1973; Munck et al., 1970). *Opaque-2* maize after years of extensive studies, failed due to poor agronomic performances and physical properties of the grain. Agricultural scientists from the International Maize and Wheat Improvement Center (CIMMYT) bred modified *opaque-2* maize populations into hard endosperm texture corn populations called quality protein maize (QPM). The QPM populations have intermediate to hard endosperm texture with high levels of lysine and tryptophan. QPM kernels contained cells tightly packed with relatively few air spaces around the starch granules, resulting in increased endosperm hardness (National Research Council, 1988). Thus, QPM can be harvested and handled without cracking or crushing of the kernels. QPM is not as susceptible to ear rot as *opaque-2* maize and

is tolerant to stored grain insects.
Commercial QPM hybrids, developed in South
Africa, have similar yields and grain
properties to their best commercial regular
corn hybrids (Gevers, 1990). Experimental QPM
hybrids in the U.S.A. also have competitive
yields and similar kernel characteristics to
commercial corns.

In 1967, Watson and Yahl determined that
soft, *opaque-2* corns had acceptable starch
yields and starch properties; but, with
significantly higher losses of soluble
proteins in the steep- and mill-waters. Van
Twisk et al (1976) found that the starch
yield of *opaque-2* corn compared favorably
with the values obtained from commercial
white corns. The amount of soluble nitrogen
lost in the steep water was almost twice as
much as in standard commercial varieties.
There is no current information on the wet
milling properties of QPM; therefore, the
purpose of this study was to compare wet
millability of QPM with regular dent corns
with contrasting endosperm textures.

MATERIALS AND METHODS

<u>Grains</u>. QPM No 387, an experimental
yellow hybrid, was grown at College Station,
TX. in 1988. Asgrow 404Y, a commercial yellow
food corn widely used by the snack food
industry, was grown in 1988 at Uvalde, TX.
Funk's G4673B, a typical yellow feed corn
currently used by the wet milling industry,
was grown at College Station, TX in 1989.

<u>Wet-milling procedure</u>. Sound corn
kernels were wet-milled according to the
procedure developed by Watson et al (1955)
with some modifications (Fig. 1).

Figure 1
Wet Milling Procedure

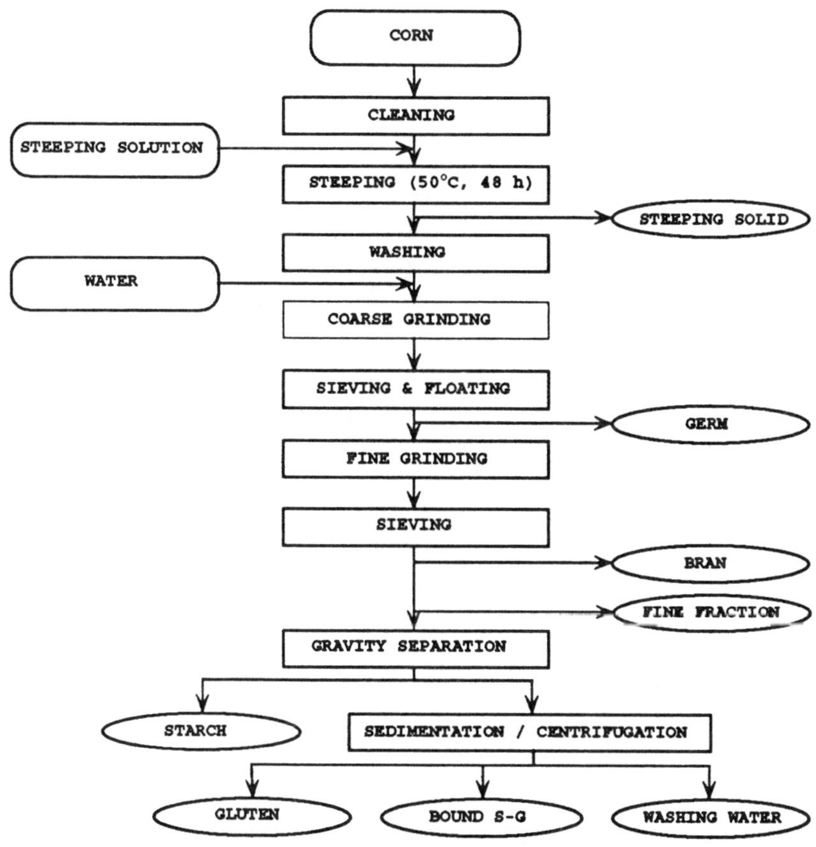

A 100 g sample of corn was mixed with 200 ml steep solution (1.48 g Na bisulfite and 4.71 ml of 85% lactic acid per liter) and placed in an oven regulated at 50°C for 48 h. After steeping, the steep solution was drained and collected for analysis.

The steeped maize was first coarsely ground with a Waring blender at low speed to free the embryos. Embryos were separated by straining through a colander. The resulting

degerminated slurry was reblended in a Waring
blender at full speed. The fine grinding
freed the endosperm from the bran and
released most of the starch granules. Fiber
was separated by straining the ground slurry
through a U.S. Standard 200 sieve and bottom
pan. The material retained on top of No 200
sieve was "fiber", and the material collected
in the pan was "gluten-starch". The gluten-
starch slurry (containing gluten, starch and
the mill solubles) was stirred and decanted
on the separation gravity table. Most of the
starch sedimented on the table. The gluten
from the protein-starch slurry was separated
according to the procedure of Steinke et al
(1991). The protein slurry was collected and
allowed to settle for 24 hr at 4°C. Most of
the slurry water was decanted, the slurry was
stirred and allowed to settle for another 24
hr at 4°C. The unpurified gluten fraction was
then transferred to 50 ml centrifuge tubes
and centrifuged at 5,000 x g for 30 min.
Thereafter, the water was decanted and
combined with the wash-water, the gluten
layer (yellowish) was hand separated from the
distinctive starch layer (whitish). Between
the two layers was an off-white colored
interphase composed of protein-bound starch.
This layer, called "bound", was carefully
collected. The starch layer was resuspended
for one additional run through the separation
table. The bound layer was also reslurried,
centrifuged, and separated as described
above. All recovered fractions were dried for
24 hr in an oven set at 50°C.

<u>Analytical methods</u>. Kernel density was
measured in a Multipycnometer(MUP-1 S/N232,
Quantachrome Corp., NY). Hardness values were
determined by subjecting kernels to abrasive
decortication for 10 min; values were
reported as a percent of the material
removed. The higher the hardness

(decortication) value, the softer the corn. Kernel weight was determined by weighing 1,000 whole kernels.

Moisture, protein (N x 6.25), fat (ether extract), starch and ash contents were determined using standard AACC (1986) procedures. Insoluble and soluble dietary fibers were determined using the colorimetric procedure of Englyst and Cummings (1988). For lysine analysis, samples were hydrolyzed in 6 N HCl for 24 h in a nitrogen rich atmosphere. The hydrolyzates were evaporated at 37°C to dryness, washed 3 times with distilled water and evaporated. Hydrolyzates were suspended in a citrate-HCl buffer adjusted to pH 2.2. Amino acids were quantified by ion-exchange chromatography after reaction with ninhydrin in a Beckman 120C Amino Analyzer (Spackman et al 1958).

Water soluble solids of corn were determined according to the procedure of Watson and Yahl (1967). Freshly ground corn (25 g) was extracted with 200 ml water in a Waring blender at full speed for 2.0 min. Then, the slurry was centrifuged for 10 min at 3,000 x g, the supernatant dried at 100°C, and expressed as percent of the original corn weight.

Pasting viscosity of isolated starch (8.6% solids) was determined using a Brabender Viscoamylograph (type VAV, Model 3042, C.W. Brabender Instruments, Inc.) equipped with a 125 cmg sensitivity cartridge, and a small amylograph cup (40 ml) rotating at 75 rpm. The amylograph cycle was set for 30 min heating from 50 to 95°C, 16 min holding at 95°C and 30 min cooling from 95 to 50°C.

Differential scanning calorimetry (DSC) of regular and QPM starches was performed using a DSC-2 Perkin-Elmer analyzer. Starch was mixed with a sufficient amount of water in a vial to give a paste containing 65 to

70% moisture. The vial was then capped and stored overnight at 4°C. After equilibration to room temperature, the starch (3-5 mg, d.b.) sample was placed into a tared pan, and tamped down with a fine glass rod to ensure even packing. The pan was sealed and weighed. Samples were analyzed at a heating rate of 10°C/min over the temperature range 40-140°C. The onset temperatures (To), the peak temperatures (Tp), and the recovery (Tr) were determined.

For X-ray diffraction, starch samples were packed in an aluminum frame, equilibrated to 96% relative humidity. and scanned with CuKα radiation, using a voltage of 35 KV, electric current of 12.5 mA, time constant 4 sec, scanning speed goniometer 1°/min and chart speed 1 cm/min.

Starch solubility at 120°C was measured by high-performance, size-exclusion chromatography (HPSEC) according to the procedure of Jackson et al (1988). Starch (0.25 g) was suspended in 100 ml of water. A 10 ml subsample was gelatinized in a pressure cooker, equilibrated, sonicated, centrifuged, filtered through a 5.0 μm nylon filter, and injected into the HPSEC system.

RESULTS AND DISCUSSION

<u>Kernel characteristics</u>. Ground food corn produced a lower amount of water solubles than QPM and feed corns (Table 1). Water solubles were positively correlated to hardness (r^2=0.97), suggesting that QPM and feed corn contained higher water solubility than food corn because QPM and feed corn had a softer kernel texture and probably less tightly packed kernel components. Watson and Yahl (1967) indicated that *opaque-2* (a high lysine corn) had a large amount of water-soluble material, probably composed of

protein, peptides, amino acids, and other
simple nitrogenous material.

Table 1
Physical Properties of Corn Kernels and
Flour

	QPM	Food Corn	Feed Corn
Moisture Content (%)	9.20±0.10	9.70±0.10	9.10±0.10
Water Soluble Solids (%)	9.09±0.01	6.82±0.16	9.12±0.04
Density (g/100 cc)	130.8±0.1	132.2±0.1	131.4±0.1
Hardness[1] (%)	42.0±0.4	33.5±0.1	44.0±0.1
1000 Kernel Weight (g)	307.0±0.7	350.4±0.7	340.2±0.7

[1] Percent material removed by abrasion.
Kernels that are softer give higher
values.

According to Watson (1984) half the
soluble solids came from the germ and the
other half from the endosperm. QPM and feed
corn were only slightly less dense than the
food corn (Table 1). Density is related to
the existence of internal pores in a particle
(Peleg, 1983), suggesting in this case that
QPM and feed corn kernels had more void
spaces or less tightly packed structure than
a regular food corn. Watson and Yahl (1967)
reported that *opaque-2* kernels had 11% lower
density than regular dent corn. Abrasive
decortication removed 42 and 44% of the QPM
and feed corn, respectively, indicating that
these two hybrids were softer than food corn.
Szaniel et al (1984) indicated that kernel
hardness was related to differences in horny-

floury ratios, pericarp thickness and cell structure. Christensen et al (1969) explained that a thick protein matrix surrounded starch granules in the horny endosperm, while a thin incomplete matrix was prevalent in floury endosperm. *Opaque-2* maize was bred to produce a higher proportion of glutelin with little zein, which gave a soft floury endosperm (Watson, 1984). The kernel size of the QPM hybrid grain was smaller than both food and feed corns. New QPM hybrids under development have characteristics comparable to those of regular dent corns (unpublished data, Cereal Quality Lab, 1990).

<u>Steeping solution uptake</u>. QPM and feed corn absorbed steep-water faster than food corn (Fig. 2).

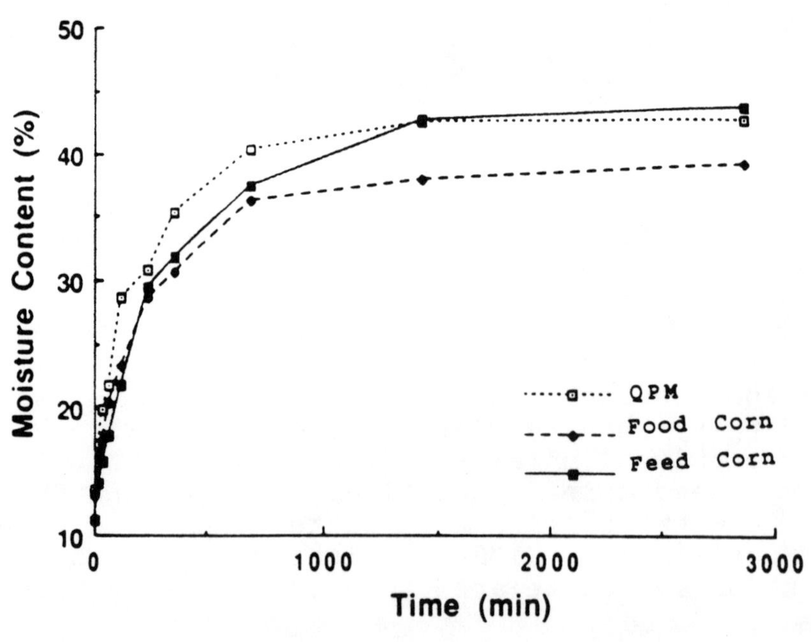

Figure 2
Steeping Solution Uptake of QPM and Regular Corns

When grain was tempered, water was presumably first held by the bran and germ and then diffused into the floury endosperm. The diffusion rate into the floury endosperm was affected by the structure of the more impermeable peripheral and corneous endosperms. A faster water penetration occurred in QPM and feed corn because they had less dense endosperm texture. Watson (1984) reported that water absorption during corn steeping causes 55 to 65% volume expansion due in part to the disruption of disulfide bonds. The protein matrix gradually swells, becomes globular and finally disperses during SO_2-steeping (Cox et al 1944).

<u>Composition of corn genotypes</u>. Interactions of plant genotype and environment conditions determine the chemical composition of corn. QPM had similar amounts of protein as the food and feed corns (Table 2). QPM protein quality was higher that of regular corn due to a better amino acid balance (Sproule et al, 1988). Starch is the major component of corn. QPM had the lowest amount of starch followed by feed and food corns. Starch content varies inversely with the protein content of endosperm; starch tends to be lower in corn with high oil content (Freeman, 1973). QPM contained a higher amount of oil than the regular corns. This is because germ size was increased for the QPM hybrid and the proportion of endosperm was reduced. QPM contained the highest amount of total dietary fiber and ash. Pericarp was the main source of both fiber and minerals. Because QPM kernels were smaller, higher fiber and ash contents were expected.

Table 2

Proximate Composition[1] of Corn Hybrids

Component	QPM	Food Corn	Feed Corn
Protein (Nx6.25) (%)	10.45±0.15	10.47±0.10	10.17±0.10
Starch (%)	78.60±0.67	81.20±0.72	79.60±0.75
Fat (%)	4.40±0.10	3.60±0.10	4.02±0.10
Dietary Fiber (%)	4.59±0.07	3.72±0.22	4.34±0.13
Ash (%)	1.48±0.16	0.83±0.06	1.26±0.12

[1] Values are expressed on dry basis.

<u>Wet milling</u>. Wet-millability of QPM was evaluated based on yields and purity of fractions (Table 3 and 4). Total fraction recovery for the three corns tested ranged from 93.9 to 96.1%. Yields of germ, fiber, starch and gluten from QPM compared favorably with those of food and feed corns. Industrial wet milling yields are much higher than those obtained at lab scale. QPM bound yield is lower than those of food and feed corns. Watson and Yahl (1967) indicated that the bound yield was a sensitive indicator of starch-gluten separation, since poor starch separation produced high bound yield. Apparently, the starch in QPM endosperm was easily released from the protein matrix. This could be an advantage for wet milling.The QPM germ retained the highest amount of protein after processing (Table 5).

Table 3
Wet Milling Yields of the Corn Fractions.

Component	QPM	Food Corn	Feed Corn
Germ			
Fraction Yield (g/100g grain)	8.6±0.8	8.7±0.4	9.0±0.7
Fiber			
Fraction Yield (g/100g grain)	14.6±1.0	13.0±0.8	16.1±0.4
Starch			
Fraction Yield (g/100g grain)	58.6±1.0	57.0±1.8	58.1±0.8
Gluten			
Fraction Yield (g/100g grain)	4.1±0.8	5.7±2.1	3.8±1.7
Bound			
Fraction Yield (g/100g bound)	1.1±0.5	3.0±2.6	2.4±0.9
Steeping Solubles	5.2±0.3	4.1±0.5	4.7±0.7
Fraction Yield (g/100g steep-soluble)			
Milling Solubles	3.2±0.3	2.5±0.3	2.4±0.6
Fraction Yield (g/100g mill-soluble)			
Recovery (Total Dry Matter) (%)	95.2±2.5	93.9±3.8	96.1±0.6

Table 4
Proximate Composition (expressed as dry matter) of the Corn Fractions.

Component	QPM	Food Corn	Feed Corn
Germ			
Protein (g/100g germ)	19.3±0.3	12.3±0.1	17.6±0.5
Fat (g/100g germ)	41.4±2.5	40.8±1.1	48.6±0.8
Ash (g/100g germ)	1.5±0.1	1.0±0.1	1.6±0.1
Fiber			
Protein (g/100g fiber)	14.3±0.3	11.9±0.3	11.7±0.2
Fat (g/100g fiber)	6.8±0.1	5.6±0.3	3.9±0.4
Starch			
Starch (g/100g starch)	>99.2	>99.2	>99.2
Protein (g/100g starch)	<0.4	<0.4	<0.4
Fat (g/100g starch)	<0.2	<0.2	<0.2
Gluten			
Protein (g/100g gluten)	38.9±1.3	39.7±0.1	49.4±0.5
Fat (g/100g gluten)	10.6±0.2	8.0±0.1	9.5±0.1
Ash (g/100g gluten)	0.7±0.1	0.3±0.1	0.9±0.1
Starch + Fiber (by diff.) (g/100g gluten)	50.2	52.0	40.2
Steep.Soluble			
Protein (g/100g s.sol.)	42.2±1.2	33.1±2.3	40.2±1.3
Mill.Soluble			
Protein (g/100g m.sol.)	53.4±1.7	44.3±1.3	46.6±1.8

Table 5
Concentration of Total Protein in Wet-
Milled Fractions

Component	Percent of Total Protein		
	QPM	Food Corn	Feed Corn
Germ	16.3	10.5	15.7
Bran	10.6	7.6	7.9
Gluten	15.3	22.0	18.7
Fine Fraction	9.7	6.7	9.8
Steep. Soluble	21.0	13.4	18.7
Mill. Soluble[1]	16.3	10.5	10.1
NC[2]	10.8	29.3	19.0

(1) Mill-Solubles includes solids from steeping and milling liquors.
(2) Not considered: it includes: the amount of protein in starch, the bound fraction plus losses during processing

This could be beneficial because the germ is superior in protein quality and quantity compared to the other anatomical parts (Bressani and Mertz, 1958; Mertz et al, 1966; Paulis and Wall, 1969). QPM fiber contained more total protein than fiber from regular corns. Pericarp, tipcap and unmilled endosperm were the main sources recovered in the fiber fraction. QPM gluten contained the lowest amount of protein but QPM protein had the best essential amino acid balance. High lysine mutants have reduced levels of zein fraction and increased levels of albumins, globulins and glutelins (Misra et al 1975). QPM steep- and mill-water had the highest amount of protein due to the increased amount of soluble proteins (albumins and globulins). This agrees with results obtained by Watson

and Yahl (1967) during the wet processing of
opaque-2 corn. Hamilton (1951) indicated
that about 60% of the germ proteins are
soluble in salt solution, while only 9% of
the endosperm proteins are soluble. Although
the steep- and mill-water are by-products
from the wet milling industry, the better
quality of QPM protein could substantially
raise the selling prices of steep- and mill-
water concentrates. QPM contained the highest
amount of fat (Table 2) likely due to the
high germ/endosperm ratio or kernel size.
Consequently, the germ, bran, and gluten
contained higher fat contents. The higher fat
content in QPM and regular corn glutens (>8%
fat) probably resulted from the disruption of
some germ cells during lab processing. Watson
(1984) reported 4% fat in commercial gluten.
Most of the oil released during wet-milling
is absorbed by the gluten. No significant
differences were found in the starch yields
of the three tested corns, suggesting that
QPM can economically replace regular corns in
wet-milling operations.
      The lysine content of grain and gluten
was highest for QPM (Table 6).

Table 6
Lysine Content (g/100 g protein) in
Kernel, Gluten and Milling soluble

Component	QPM	Food Corn	Feed Corn
Whole Kernel	4.11±0.10	3.15±0.09	3.34±0.10
Gluten	3.06±0.15	1.26±0.15	2.43±0.12
Milling soluble	3.00±0.12	2.71±0.05	3.41±0.09

The amino acid scores of QPM, food and
feed corns, based on a requirement of 5.44 g
lysine/100 g protein, were 79.0, 59.0,and
62.0, respectively. Gluten is the prime by-
product of the wet milling industry. Almost
all of the corn gluten produced by wet-
milling is utilized in animal feeds. A gluten
with a higher lysine content can be
beneficial in diet formulation for swine and
poultry since it can significantly reduce use
of expensive protein supplements.

More than 80 % of the triglyceride
composition of all corn oils, was comprised
of unsaturated fatty acids (oleic and
linoleic acids) (Table 7).

Table 7
Concentration of fatty acids in corn oil
triglycerides

Fatty Acid	Percent of Total Fatty Acids in the Oil		
	QPM	Food Corn	Feed Corn
Palmitic (16:0)	14.16	12.20	13.22
Palmitoleic (16:1)	0.27	0.28	0.25
Stearic (18:0)	2.40	2.31	2.23
Oleic (18:1)	37.87	45.65	34.35
Linoleic (18:2)	44.73	38.98	49.32
Linolenic (18:3)	0.57	0.58	0.63

QPM contained more palmitic acid than
regular corns; the food corn contained more
oleic acid than QPM and feed corn; and the
feed corn contained more linoleic acid than
QPM and food corn.

Starch is the prime product from corn
wet-milling. No significant differences were
found in Brabender viscosities between QPM
and regular corns (Fig. 3). All samples
developed the typical pasting properties of
regular maize starch (Greenwood, 1978).
Starches solubilized at 120°C had similar
amylopectin:amylose ratio and average
apparent molecular weights (Table 8).

Figure 3
Amylograph Viscosity of QPM and Regular Corns

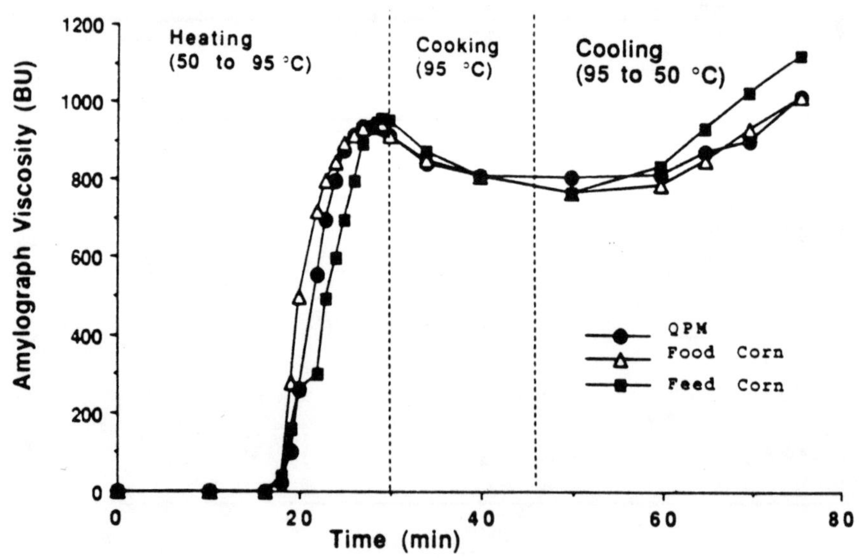

Table 8
Properties of Isolated Starches Obtained by High
Performance, Size-Exclusion Chromatography.

Starch	Soluble Starch (%)	AMP:AMY Ratio	AMP-MW $(x10^7)$	AMY-MW $(x10^5)$
QPM	75.5±0.3	70.5:29.5	1.71	2.23
Food Corn	71.6±0.4	70.0:30.0	1.71	1.84
Feed Corn	76.1±0.6	69.5:30.5	1.72	2.63

Values are means ± standard deviation. Molecular
weight of amylopectin and amylose are
expressed as average MW.

The thermal properties of the three
starches were nearly identical and similar to
those reported for laboratory isolated corn
starch (Krueger et al 1987) (Table 9).

Table 9
Differential Scanning Calorimetry
Characteristics of Isolated Starches.

Starch	Peak Start ($^{\circ}$C)	Onset Temp. ($^{\circ}$C)	Endotherm Min./Max. ($^{\circ}$C)	Peak End ($^{\circ}$C)
QPM	71.40	71.92	74.55	79.37
Food Corn	70.77	71.87	74.80	80.47
Feed Corn	70.15	70.75	73.62	79.53

No significant differences in the
crystallinity patterns was found when QPM

starch was compared with food and feed corn
starches (Fig. 4). As expected, all starches
had the typical A-type structure with a
strong intensity at 5.78, 5.17, 4.86, and
3.78 Å (Zobel, 1964). Watson and Yahl (1967)
reported that the starch recovered from the
opaque-2 corn was similar in all properties
to regular corn starch. The similarity among
starches indicate that the isolated starch of
QPM grain can be used for production of
starch-based products.

Figure 4
X-Ray Patterns of QPM and Regular Corn
Starches

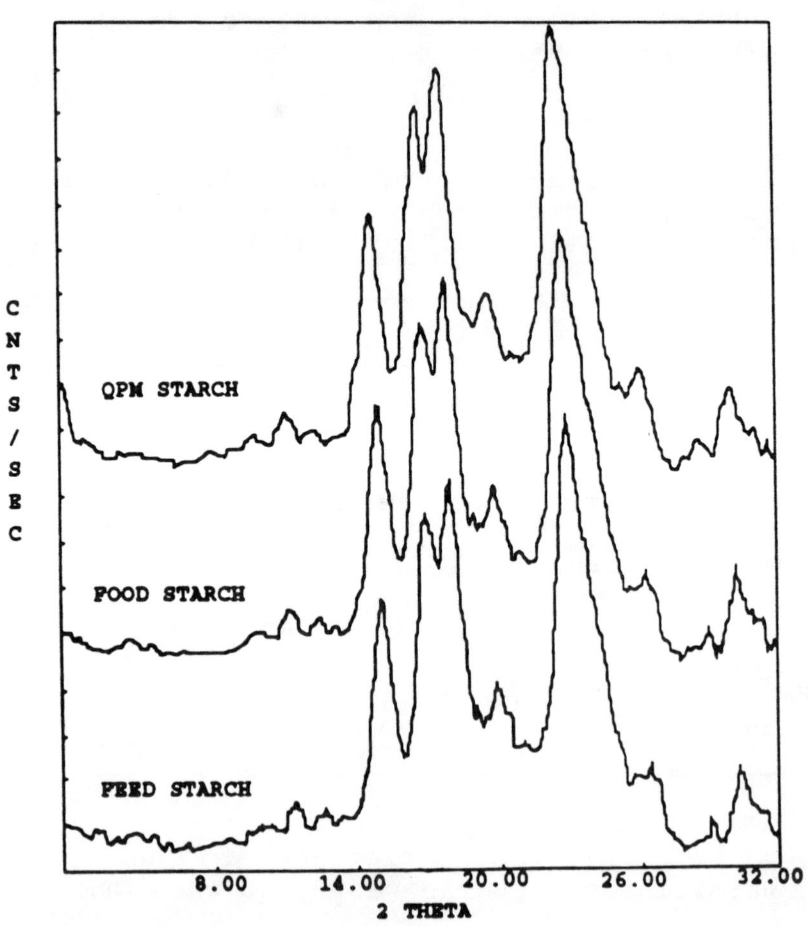

CONCLUSIONS

QPM yielded similar amounts of the major wet-milling products comparable to feed and food corn hybrids. QPM produces gluten with enhanced lysine composition and greater value for swine and poultry feeds. QPM oil had a fatty acid composition similar to regular corns. QPM starch had viscoamylograph and gelatinization properties similar to regular corn starches.

ACKNOWLEDGEMENT

We acknowledge partial support for this research from the Snack Foods Association, Alexandria, VA.

Contribution TA 30299 from the Texas Agricultural Experiment Station, Texas A&M University, College Station, TX 77843-2474.

LITERATURE CITED

AACC. 1986. "Approved Methods". American Association of Cereal Chemists, St.Paul, MN.

Bressani, R. and Mertz, E.T. 1958. Studies on corn proteins. IV. Protein and amino acid content of different corn varieties. Cereal. Chem. 35:227-235.

Bockholt, A.J. Current status of QPM hybrids for the U.S. Presented at the Quality Protein Maize (QPM) Symposium. American Association of Cereal Chemists. 1990 Annual Meeting. October 14-18. Dallas, TX.

Christensen, D.D., Nielsen, H.,C., Khoo, U., Wolf, M.J., and Wall, J.S. 1969. Isolation and chemical composition of protein bodies

and matrix proteins in corn endosperm.
Cereal Chem. 46:372-381.

Cox, M.J., Mac Master, M.M., and Hilbert,
G.E. 1944. Effect of the sulfurous acid
steep in corn wet-milling. Cereal Chem.
21:447-465.

Englyst, H.N. and Cummings, J.H. 1988. An
improved method for measurement of dietary
fibre as the non-starch polysaccharides in
plant foods. J. Assoc. Off. Analyt. Chem.
71, 808-814.

Freeman, J.E. 1973. Quality factors affecting
value of corn for wet milling. Trans. ASAE
16:671-679.

Greenwood, C.T. 1978. Starch. Page 135 in:
Advances in Cereal Science and Technology.
Volume II. Y. Pomeranz, ed. Am. Assoc. of
Cereal Chem.:St. Paul, MN.

Hamilton, T.H., Hamilton, B.C.,Johnson, B.C.,
and Mitchell. 1951. Cereal Chem. 28:163.

Jackson, D.S., Choto-Owen, C., Waniska, R.D.
and Rooney, L.W. 1988. Characterization of
starch cooked in alkali by aqueous HPLC-SEC.
Cereal Chem. 65(6):493-496.

Krueger, B.R., Knutson, C.A., Inglett, G.E.,
and Walker, C.E. 1987. A differential
scanning calorimetry study on the effect of
annealing on gelatinization behavior of corn
starch. J. Food Science 52:715.

Mertz, E.T., Nelson, O.E., Bates, L.S., and
Veron, O.A. 1966. Better protein quality in
maize. In: World Protein Source. R.F. Gould
(ed.). Adv. Chem. Ser. 56:228-242.

Mertz, E.T., Bates, L.S., and Nelson, O.E.
1964. Mutant gene that changes protein
composition and increases lysine content of
maize endosperm. Science 145:279-280.

Misra, P.S, Mertz, E.T., and Glover, D.V.
1975. Studies on corn proteins. Endosperm
changes in single and double endosperm
mutants of maize. Cereal Chem. 52:161-166.

Munck, L., Karlson, K., Haldberg, M. and
Eggum, B. 1970. Gene for improved

nutritional value in barley seed protein.
Science 168:985.

National Research Council. 1988. Quality
Protein Maize. National Academic Press,
Washington, D.C.

Paulis, J.W. and Wall, J.S. 1969. Albumins
and globulins in extracts of corn grain
parts. Cereal Chem. 46:263-273.

Peleg. M. 1983. Physical Characteristics of
Food Powders. In: Physical Properties of
Foods. M. Peleg and E.B. Bagley (Eds.). AVI
publishing Company, Inc. Westport,
Conneticut. Page: 293.

Singh, R. and Axtell, J.D. 1973. High lysine
mutant gene that improves protein quality
and biological value of grain sorghum. Crop
Sci. 13:535.

Spackman, D.H., Stein, W.H., and Moore, S.
1958. Automatic recording apparatus for use
in chromatography of amino acids. Anal.
Chem. 30:1190.

Sproule, S., Serna-Saldivar, S.O. Bockholt,
A.J., Rooney, L.W., and Knabe, D.A. 1988.
Nutritional evaluation of tortillas and
tortilla chips from quality protein maize.
Cereal Foods World. 33(2):233-236.

Steinke, D.J. and Johnson, L.A. 1991.
Steeping of maize in the presence of
multiple enzymes. I. Static batchwise
steeping. Cereal Chem.(In press)

Szaniel, J., Sagi, F., and Palvolgyi, I.,
1984. Hardness determination and quality
prediction of maize kernels by a new
instrument, the molograph. Maydica 29:9-20.

Van Twisk, P., Quicke, V., and Gevers, H.O.
1976. Physical properties and biological
evaluation of high lysine maize. Cereal
Chem, 53:692.

Watson, S.A., Hirata, Y., and Williams, C.B.
1955. A study of the lactic acid
fermentation in commercial corn steeping.
Cereal Chem. 32:382.

Watson, S.A., and Yahl, K.R. 1967. Comparison of the wet-milling properties of *opaque-2* high lysine corn and normal corn. Cereal Chem. 44:488-498.

Watson, S.A. 1984. Corn and sorghum starches: Production. Pages 417-468 in:"Corn: Chemistry and Technology". R.L. Whistler and E.F., Paschal, eds. Academic Press: New York.

Zobel, H. 1964. X-ray Analysis of Starch Granules. Page 109. In: Methods in Carbohydrate Chemistry. Vol. 4.R.L. Whistler (Ed.). Academic Press. New York.

DRY MILLING OF QUALITY PROTEIN MAIZE

Y. V. Wu
National Center for Agricultural
Utilization Research,
Agricultural Research Service,
U.S. Department of Agriculture
Peoria, IL 61604

INTRODUCTION

Ordinary maize (corn) is deficient in lysine and tryptophan for human consumption. Mertz et al (1964) found one mutant maize (opaque-2) that had about twice the normal levels of lysine and tryptophan. Opaque-2 maize had soft opaque kernels, however, instead of hard, transparent ones typical of most maize grown worldwide. Thus, upon dry milling, opaque-2 maize produced poor yields of grits compared with ordinary maize. Opaque-2 maize also yielded less grain than ordinary maize, and was more subject to fungal and insect infestations.

Breeders at the International Maize and Wheat Improvement Center (CIMMYT), Mexico, subsequently transformed opaque-2 maize into genotypes with high nutritional quality, high yield, normal moisture content, traditional appearance, and conventional hardness (NRC, 1988). These new varieties are called "quality protein maize" (QPM). Several U.S. research groups are now using these lines to produce commercial QPM hybrids suitable for local growing conditions. We here report dry milling results of QPM from CIMMYT, experimental QPM varieties and hybrids from the United States, and conventional dent corn.

MATERIALS AND METHODS

Quality Protein Maizes

White dent-1 QPM came from CIMMYT in 1988. F337 is an open pollinated, experimental QPM grown in Texas in 1988, originally derived from Population 70, temperate yellow dent QPM of CIMMYT. Experimental hybrid 1 is a modified endosperm opaque-2 corn, not completely modified (some kernels are still opaque). Synthetic disease oil and Elite Single Cross are synthetic varieties for increased endosperm modification. The last three genotypes are experimental lines from Illinois.

Commercial dent, DeKalb 615F, and IH 119 X IH51 are three normal commercial dent corns from Illinois commercial sources.

Dry Milling

Corn (4,500 g) was tempered from initial moisture content to 16% for 16 hr, then to 21% for 1.75 hr, and to 24% for 15 min before degerming. All tempering was at room temperature (24-26°C). Corn with added water was rotated slowly in a stainless steel drum for about 15 min, and then maintained at room temperature. Tempered corn was processed at 1650 rpm in an experimental horizontal drum degermer equipped with a steel screen with 7.9 mm diameter round holes, as described by Brekke et al (1971).

Degermer stock was spread out and allowed to dry to 17± 1% moisture at room temperature. Three pairs of corrugated rolls, a Bates laboratory aspirator, and a box sifter holding three 30 x 30 cm sieves were used to process the degermer stock into degerminator fines; low fat meal; low fat flour; high fat meal; high fat flour; first, second, and third break grits; bran meal; first, second, and third germs; and hull fractions (Figure 1).

Density

A Beckman (Fullerton, CA) Air Comparison Pycnometer Model 930 was used to measure the volume

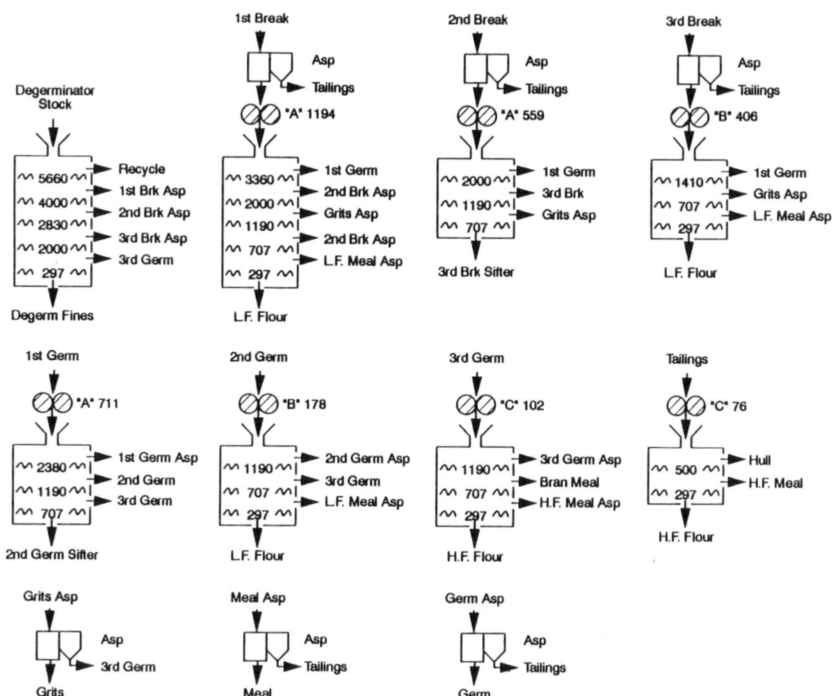

Fig. 1. Roller-milling flow sheet. Degerminator stock from tempered corn was milled into 1st break grits, 2nd break grits, 3rd break grits, high-fat meal, low-fat meal, high-fat flour, low-fat flour, 1st germ, 2nd germ, 3rd germ, bran meal, and hull. Brk = break; Asp = aspirator; LF = low-fat; HF = high-fat, "A", "B", and "C" are roller mill clearance setting in μm. $\sim$ are sieves with square opening in μm. Adapted from Peplinski et al, 1984.

(about 26 ml) of maize kernels in a cup from the pycnometer. The weight of maize kernels was determined with an analytical balance. Maize density was determined by dividing mass of kernels (grams) by volume (ml). Maize was equilibrated to about 11% moisture in a constant temperature and constant humidity room before measurement. Change in density with moisture content from 5 to 14% for DeKalb 615F was determined, and the density of each maize was adjusted to 11% moisture based on the relationship observed for DeKalb 615F. Lower moisture values for grain were obtained by partial drying in an air oven at different temperatures. Higher moisture values for grain resulted from grain kept in desiccators above saturated salt solutions with higher relative humidity. Density was determined in triplicate.

Analyses

Ground maize samples for amino acid analyses were hydrolyzed with 6N HCl for 4 hr at 145°C (Gehrke et al., 1987). Cystine and methionine were oxidized with performic acid prior to hydrolysis (Moore, 1963). Amino acids were quantified by cation exchange chromatography using a Beckman 6300 amino acid analyzer (Beckman Instruments, Inc., San Raman, CA). Tryptophan was determined by colorimetric detection after enzymatic hydrolysis by pronase (Spies and Chambers, 1949; Holz, 1972).

Moisture of duplicate dry-milled fractions was measured by drying in an air oven at 135°C for 2 hr (AACC, 1983). Moisture content for tempering corn was determined in triplicate by a Brabender Moisture/Volatiles Tester, type SAS (C. W. Brabender Instruments, Inc., Hackensack, NJ) after corn was cracked in an Enterprise Model 00 grain mill (Philadelphia, PA). Nitrogen, fat, ash, and crude fiber contents were determined by AACC methods (1983). Protein was estimated from N x 6.25 (Drake et al., 1989); Baudet et al (1986) reported a nitrogen to protein conversion factor of 5.77 for maize, but this value has not yet been generally accepted. Fatty acid composition was determined by gas chromatography (Wilcox et al., 1984). Starch was determined by polarimetry (Garcia and Wolf, 1972).

In statistical analysis, p is probability and LSD is least significant difference.

RESULTS AND DISCUSSION

Proximate Composition

Starch, protein, ash, crude fiber, fat, and fatty acid compositions of five QPM and three normal dent corns are listed in Table I. A general linear model procedure was used to compare the five QPM samples as a group with the three normal dent corns as a group. The normal dent corns had higher starch (p = 0.008, LSD = 2.6), lower fat (p = 0.007, LSD = 0.8), lower percent 18:1 fatty acids (p = 0.0001, LSD = 3.5), higher percent 18:2 fatty acids (p = 0.002, LSD = 4.9), and higher percent 18:3 fatty acids (p = 0.027, LSD = 0.25) than QPM. There was no significant difference (p > 0.05) between normal dent corns and QPM for protein, ash, crude fiber, percent 16:0, and percent 18:0 fatty acids.

Dry Milling

Yields of dry-milled fractions from five QPM samples were compared with those from three normal dent corns (Table II). Yields of total grits and prime products (total grits + low fat meal + low fat flour) for Elite S. C. and Exp. Hybrid 1 were lower than from other QPMs (Table II). Yields of total grits and prime products from five QPM covered a range that included three conventional dent corns. Degerming and roller milling of QPM was normal.

Density

Change in density of DeKalb 615F corn (moisture content 5.5-14%) can be described by the linear equation (correlation coefficient -0.977, p < 0.01): Percent moisture = 471.3 - 346.4 density. This equation predicts that density of DeKalb 615F corn decreases by 0.0029 unit when moisture increases from 11 to 12%. Based upon this relationship, density

TABLE I
Composition of Quality Protein Maize and Commercial Dent Corns (% Dry Basis)[a]

	Starch	Fat	Fatty Acid Composition				
			16:0	18:0	18:1	18:2	18:3
QPM							
White Dent-1	68.3	5.2	14.1	1.1	36.3	47.9	0.6
F337	69.3	5.4	13.8	1.4	37.6	46.8	0.4
Elite Single Cross	69.4	5.0	10.7	1.2	33.6	54.0	0.5
Experimental Hybrid 1	69.7	5.1	11.2	1.0	35.1	52.0	0.7
Synthetic Disease Oil	65.6	6.3	11.7	1.3	38.4	48.3	0.3
Normal							
Commercial Dent	73.0	4.1	14.2	1.4	25.5	58.0	0.9
DeKalb 615F	71.6	4.3	14.4	1.2	22.6	61.0	0.7
LH 119 x LH 51	73.3	3.8	14.4	1.8	21.5	61.5	0.8

[a]Ash contents of 5 QPM are 1.4, 1.7, 1.6, 1.4 and 1.6% vs 1.2, 1.5 and 1.0% for 3 normal corns. Crude fiber contents of 5 QPM are 4.1, 3.5, 3.3, 4.1 and 3.5% vs 3.3, 4.2 and 2.9% for 3 normal corns. Protein contents of 5 QPM are 10.6, 10.5, 10.0, 9.9 and 10.9% vs 9.1, 10.1 and 9.9% for 3 normal corns.

TABLE II
Yields of Dry Milling Fractions from Quality Protein Maize and Commercial Dent Corns

	Deg. Fines	L. F. Meal	L. F. Flour	1st Break Grits	2nd + 3rd Break Grits	H. F. Meal	H. F. Flour	Bran Meal	Hull	1st + 2nd + 3rd Germs	Density g/cc
White Dent-1	0.6	10.4	2.9	15.7	39.7	5.8	3.0	2.7	5.0	14.2	1.3272
F337	1.3	10.6	4.1	16.1	34.9	6.8	2.6	2.9	5.1	15.6	1.3178
Elite S. C.	1.8	11.3	4.2	4.7	35.6	12.6	5.4	3.3	6.4	14.8	1.2971
Exp. Hybrid 1	2.8	12.9	6.3	6.8	32.3	9.6	6.3	2.2	5.6	15.1	1.2613
Synthetic D.O.	1.1	11.4	3.2	8.3	40.5	10.2	2.8	4.0	4.6	14.0	1.3217
Commercial Dent	2.3	10.8	4.3	10.3	34.6	6.4	1.2	8.3	6.0	15.7	1.3027
DeKalb 615F	1.7	9.5	3.2	18.6	33.7	7.5	2.8	3.2	5.7	14.0	1.3305
LH 119 X LH 51	1.1	10.1	3.2	19.2	35.0	6.3	1.8	3.1	3.8	15.6	1.3427

S. C. = single cross; D. O. = disease oil; Deg = degerminator; L. F. = low fat; H. F. = high fat.
Prime Products = low fat meal + low fat flour + 1st-3rd break grits; the values for 5 QPM are 68.7, 65.7, 55.8, 58.3 and 63.4 vs 60.0, 65.1 and 68.5 for 3 normal corns.
First five are quality protein maize, last three are normal dent corns. Density corrected to 11% moisture.
Yields of 1st-3rd break grits are 55.4, 51.0, 40.3, 39.1 and 48.8 for 5 QPM vs 44.9, 52.3 and 55.2 for 3 normal corns.
Average standard deviations from duplicate commercial dent fractions, triplicate DeKalb 615F fractions and triplicate density of corn kernel are 0.27, 0.40 and 0.0006, respectively.

values for all corns (Table II) were adjusted to 11% moisture. Corn density positively correlated with hardness (Pomeranz et al., 1984). Hardness values of the five QPM samples (except of Exp. hybrid 1), based on density, were comparable to those of the three dent corns (Table II).

Amino Acid Composition

Amino acid compositions of five QPMs, a normal dent corn, and average values for corn are listed in Table III. Lysine content of QPM (Table III) ranged from 3.7 to 4.3 g/16 g nitrogen, compared with 3.2 for DeKalb 615F and 2.8 for an average of 101 corn samples (Drake et al., 1989). Since lysine is the first limiting amino acid in corn, the large increase in lysine content of QPM represents a significant increase in protein quality. Tryptophan is the second limiting amino acid in corn. Three of five QPMs (Table III) also had higher tryptophan contents than an average of 16 corn samples (Drake et al., 1989). In general, QPMs have higher aspartic acid + asparagine, proline, glycine, histidine, arginine, and half-cystine, but lower serine, glutamic acid + glutamine, alanine, leucine, and tyrosine than average values for dent corn (Table III).

Relationships Among Yields of Dry-Milled Fractions

Density (hardness) of QPM was positively correlated with yield of total grits ($p < 0.05$), but negatively correlated with yields of degerminator fines, low fat meal, low fat flour, high fat flour, and high + low fat flours ($p < 0.01$, $p < 0.05$, $p < 0.01$, $p < 0.05$, and $p < 0.01$, respectively) (Table IV). Total grits yield correlated positively with yields of first break grits and of prime products ($p < 0.05$ and $p < 0.01$, respectively), but negatively with yields of degerminator fines, high fat flour, and high + low fat flours (all $p < 0.05$). Other significant correlations among yields of dry-milled fractions also exist (Table IV).

TABLE III

Amino Acid Composition of Quality Protein Maizes
and Commercial Dent Corns (g/16 g N)

Amino Acids[a]	White Dent-1	F337	Elite Single Cross	Experimental Hybrid 1	Synthetic Disease Oil	DeKalb 615F	Corn[b]
Aspartic Acid	7.9	8.1	7.6	8.4	8.4	7.3	7.0
Threonine	3.9	3.9	3.7	3.8	3.9	3.7	3.8
Serine	4.0	4.2	4.1	4.1	4.1	4.2	4.7
Glutamic Acid	17.1	16.6	16.6	16.5	17.2	19.8	18.8
Proline	10.9	9.9	10.0	10.2	11.1	9.6	8.7
Glycine	4.8	4.8	4.4	4.8	4.7	4.0	4.1
Alanine	6.3	6.2	6.3	6.2	6.5	8.1	7.5
Valine	5.5	5.5	5.1	5.6	5.4	5.1	5.1
Isoleucine	3.4	3.4	3.3	3.4	3.3	3.8	3.6
Leucine	10.1	9.3	10.2	9.8	10.2	13.5	12.3
Tyrosine	2.8	2.7	2.8	2.9	2.8	3.1	4.1
Phenylalanine	4.4	4.4	4.4	4.6	4.5	5.5	4.9
Histidine	4.3	4.2	3.8	4.1	4.2	3.2	3.0
Lysine	4.0	4.3	3.7	4.0	4.0	3.2	2.8
Arginine	6.2	6.6	5.8	6.3	6.2	5.1	5.0
Half-Cystine	3.0	3.0	2.9	3.1	3.1	2.5	1.8
Methionine	1.8	1.9	1.9	1.8	1.8	2.5	2.1
Tryptophan	1.0	1.0	0.7	0.7	0.9	0.9	0.7

[a] Glutamic acid includes glutamine. Aspartic acid includes asparagine.
[b] Average of 16 to 101 samples for individual amino acids (Drake et al., 1989).

TABLE IV
Correlation Coefficients of Dry-Milled Fractions Yields from Five Quality Protein Maizes[a]

	Deg. Fines	Low Fat Meal	Low Fat Flour	1st Break Grits	High Fat Flour	Total Grits
Low fat meal	0.908*	...				
Low fat flour	0.966**	0.879*	...			
2nd & 3rd break grits	-0.869	-0.621	-0.911*	0.276		
High fat meal	0.516	0.485	0.281	-0.951*		
High fat flour	0.905*	0.811	0.823	-0.757	...	
Prime products	-0.796	-0.673	-0.617	0.924*	-0.852	
Total grits	-0.917*	-0.820	-0.783	0.883*	-0.917*	...
Density	-0.983**	-0.924*	-0.965**	0.632	-0.935*	0.878*
High & low fat flours	0.975**	0.879*	0.942*	-0.653	0.966**	-0.899*

[a]*p < 0.05, ** p < 0.01.

Prime products = total grits + low fat meal + low fat flour.

Correlation coefficients of prime products vs high fat meal, prime products vs total grits and density vs high & low fat flours are -0.908*, 0.970** and -0.992**, respectively.

CONCLUSIONS

Normal dent corns contain more starch and have a higher percentage of 18:2 and 18:3 fatty acids than QPM, but have less fat and 18:1 fatty acids. QPM can be degermed and roller milled normally. Yields of total grits and prime products from QPM are comparable to those from conventional dent corns. Thus, since QPM has higher lysine and tryptophan contents than normal dent corns, dry milling of QPM can yield products with improved nutritional value.

ACKNOWLEDGMENTS

I thank B. D. Deadmond for technical assistance; R. J. Lambert for experimental hybrid 1, synthetic disease oil, and elite single cross samples; L. W. Rooney for F337; and M. Bjarnason for white dent-1 QPM.

LITERATURE CITED

AMERICAN ASSOCIATION OF CEREAL CHEMISTS. 1983. Approved Methods of the AACC. 8th ed. Methods 44-19, 08-03, 30-26, and 32-15, approved April 1961; Method 46-13, approved October 1976. The Association: St. Paul, MN.

BAUDET, J., HUET, J.-C., and MOSSE, J. 1986. Variability and relationships among amino acids and nitrogen in maize grains. J. Agric. Food Chem. 34:365.

BREKKE, O. L., GRIFFIN, E. L., Jr., and BROOKS, P. 1971. Dry-milling of opaque-2 (high lysine) corn. Cereal Chem. 48:499.

DRAKE, D. L., GEBHARDT, S. E., and MATTHEWS, R. H. 1989. In "Composition of Foods: Cereal Grains and Pasta." U.S. Dept. of Agric., Agric. Handbook No. 8-20, p. 31. U.S. Government Printing Office: Washington, D. C.

GARCIA, W. J., and WOLF, M. J. 1972. Polarimetric determination of starch in corn with dimethyl sulfoxide as a solvent. Cereal Chem. 49:298.

GEHRKE, C. W., REXROAD, P. R., SCHISLA, R. M., ABSHEER, J. S., and ZUMWALT, R. W. 1987. Quantitative analysis of cystine, methionine, lysine and nine other amino acids by a single oxidation-4 hour hydrolysis method. J. Assoc. Off. Anal. Chem. 70:171.

HOLZ, F. 1972. Automatic determination of tryptophan in proteins and protein-containing plant products with dimethylaminocinnamaldehyde. Landwirt. Forsch., Sonderh. 27:96.

MERTZ, E. T., BATES, L. S., and NELSON, O. E. 1964. Mutant gene that changes protein composition and increases lysine content of maize endosperm. Science 145:279.

MOORE, S. 1963. On the determination of cystine as cysteic acid. J. Biol. Chem. 238:235.

NATIONAL RESEARCH COUNCIL. 1988. Quality Protein Maize. National Academy Press, Washington, D. C.

PEPLINSKI, A. J., ANDERSON, R. A., and ECKHOFF, S. R. 1984. A dry-milling evaluation of trickle sulfur dioxide-treated corn. Cereal Chem. 61:289.

POMERANZ, Y., MARTIN, C. R., TRAYLOR, D. D., and LAI, F. S. 1984. Corn hardness determination. Cereal Chem. 61:147.

SPIES, J. R., and CHAMBERS, D. C. 1949. Chemical determination of tryptophan in proteins. Anal. Chem. 21:1249.

WILCOX, J. R., CAVINS, J. F., and NIELSEN, N. C. 1984. Genetic alteration of soybean oil composition by a chemical mutagen. J. Am. Oil Chem. Soc. 61:97.

＃ THE ALKALINE PROCESSING PROPERTIES OF QUALITY
PROTEIN MAIZE

S.O. Serna-Saldivar, M.H. Gomez, A.R. Islas-
Rubio, A.J. Bockholt, and L.W. Rooney
Cereal Quality Lab.
Dept. Soil & Crop Sciences
Texas A&M University,
College Station, TX 77843-2474

INTRODUCTION

Maize (corn) is the staple food for
millions of people around the world. This
cereal is especially important in Central
America, Mexico, and some countries of Africa
and Asia where it is the main source of energy
and protein among the general population. Corn
is consumed in different ways throughout the
world, ranging from immature grain on the cob
to products processed by different traditional
technologies, i.e., nixtamalization or
alkaline-cooking for tortilla production,
fermentation for ogi production, dry milling
for thin or thick porridges. Corn, like most
cereals, has relatively low protein content
and has limiting amounts of two essential
amino acids, lysine and tryptophan. This amino
acid imbalance is relevant in maize-consuming
countries because the basic diet fails to
provide sufficient essential amino acids.
Agricultural scientists have worked
extensively to improve the quality of maize
protein. These efforts have resulted in the
development of high lysine or *opaque-2* maize

(Mertz et al., 1964). CIMMYT (International Maize and Wheat Improvement Center) transformed *opaque-2* corn into nutritionally improved harder kernel types called quality protein maize (QPM) (Vasal et al., 1980). QPM varieties and lines with white or yellow endosperm and with yield potential equal to commercial corn lines or varieties have been released in Guatemala, Paraguay, Ecuador, Brazil, China, and Africa. Recently, collaborative efforts among CIMMYT, South Africa and Texas A&M University have resulted in the experimental production of QPM hybrids with grain yields sufficiently high to successfully compete for USA food corn markets (Bockholt, 1990).

Corn is consumed as tortillas by a large group of people in Latin America. Tortillas are made by cooking the grain in water:lime solution. Cooked corn is steeped overnight and washed to produce nixtamal, which is ground into masa. Masa is shaped into thin circles and baked into tortillas (Serna-Saldivar et al., 1990).

This article reviews current information on the alkaline-cooking properties of QPM. In addition, pilot plant trials to evaluate the alkaline processing properties of yellow and white QPM hybrids compared to a commercial food grade corn hybrid are reported here.

<u>Alkaline-cooking properties of QPM</u>. One of the first studies on the use of QPM in tortilla processing was conducted in Guatemala in 1976 (Valverde et al., 1983). QPM production (soft and hard endosperm maizes), storage conditions, tortilla production by housewives, nutritional evaluation, and acceptability by families were evaluated. Among the conclusions, the agronomic data indicated that with conventional cultural practices, QPM yields were as high as those of local cultivars. QPM tortillas had

acceptability similar to those prepared from
normal corn. The replacement of normal corn by
a high-protein-quality maize, i.e. QPM,
resulted in an improvement in children's
growth rates.

Ortega et al. (1986) compared protein
differences (lysine and tryptophan) between
QPM and its normal maize counterpart during
the different stages in tortilla processing.
The study reported that QPM had 80% more
tryptophan and 64% more lysine than its normal
counterpart. Loss of total lysine during
processing was minimal for both genotypes,
however, processing reduced the tryptophan
levels by 15% in QPM and 11% in normal corn.
Similar changes in the solubility of the
protein fractions (albumins and globulins,
true zein, zein like, glutelin like, and true
glutelins) occurred during tortilla processing
for both maizes. During processing, chemical
changes (hydrophobic interactions, cross-
linking, and denaturation) occurred in the
proteins which changed the native solubility
and reduced their *in-vitro* pepsin
digestibility. Processing had no major effect
on the content of the amino acids.

The characteristics of alkaline
processing and nutritional quality of
tortillas and tortilla chips made from a QPM
hybrid and a food grade corn were evaluated by
Sproule et al. (1988). QPM required longer
cooking time because it had a greater amount
of corneous endosperm than the food grade
corn. However, QPM had a lower dry matter loss
because the pericarp was not completely
removed during cooking and steeping. Tortillas
from both QPM and food grade corn were
flexible, with good overall appearance. The
protein content of both products was similar.
QPM had more than twice the level of the
albumin/globulin fraction than did food grade
corn. Lysine was the first limiting amino acid
of both corn products (Table I), QPM and food

grade corn products provided about 70% and 50% of the lysine requirement.

   Nutritional value of QPM tortillas. The average feed intake and average daily gain of rats fed QPM products were significantly higher than for rats fed food grade corn products. Both corn lines had similar dry matter, protein and energy digestibilities (Table 1), which indicated that the increased lysine levels in QPM were responsible for the improved rat performance. In both maize types, processing grain into tortillas and tortilla chips slightly decreased protein digestibility and rat performance.

TABLE 1
Nutritional values of Asgrow 404Y and QPM

	Asgrow 404Y			QPM		
Value	Grain	Tort.	Chip	Grain	Tort.	Chip
	Amino Acid Composition (g/16 g N)					
Lysine	2.8	2.8	2.6	4.2	3.8	3.8
Trypt.	0.6	0.5	---	1.0	0.9	---
AA Score [%]	51	50	48	77	70	70
	Apparent Digestibility [%]					
Energy	92.9	94.3	94.7	90.8	94.0	94.4
Protein	86.1	83.8	82.1	85.5	80.9	81.9

Taken from Sproule et al., 1988.

   Effect of processing on protein fractionation. Vivas et al. (1990) reported changes in protein fractions, electrophoretic patterns and protein digestibility of a QPM

hybrid and a food grade maize hybrid during
tortilla and tortilla chip production. QPM had
a larger albumin/globulin fraction than food
grade corn (Table 2). Processing raw grains
into tortillas and tortilla chips reduced the
solubility of the albumin/globulin, prolamin ,
and glutelin fractions in both grains. Cross-
linked prolamins and residue proteins
increased in both corns after processing.

Table 2
Protein Solubility Distribution of QPM and
Food Grade Corn (FGC) Processed into
Tortillas and Tortilla Chips[a].

Sample	N (%)	F[b]I (%)	FII (%)	FIII (%)	FIV (%)	Res (%)
QPM						
Raw	1.79	38.8	12.3	25.8	19.1	4.6
Tortilla	1.75	12.6	12.0	27.6	17.2	17.6
Chip	1.78	11.2	10.3	27.6	10.8	28.8
FGC						
Raw	1.60	30.4	28.0	21.7	21.0	2.7
Tortilla	1.62	8.5	9.2	42.7	8.7	21.0
Chip	1.70	8.0	7.0	44.6	4.9	30.9
LSD (P<.01)		0.89	1.1	1.26	0.92	1.26

(a) Modified from Vivas-Rodriguez et al.(1990)
(b) FI: albumins/globulins; FII: prolamins;
FIII: cross-linked prolamins; FIV:
glutelins; and Res: residue.

The number and intensity of protein
bands in electrophoretic patterns decreased
after processing for both hybrids. *In-vitro*
pepsin digestibility also decreased after
processing for both grains. However, QPM
contained more albumins and globulins, and

less prolamins which accounts for the improved protein quality of QPM products.

<u>Effect of processing on Ca availability</u>. Serna-Saldivar et al (1991, 1992) evaluated the effect of lime treatment on the availability of calcium in diets of tortillas and beans. In this study, QPM, regular corn and sorghum were processed into tortillas. Weanling rats were fed raw grain or tortilla-based diets supplemented with dry, cooked pinto beans and a Ca-free or Ca-rich mineral premix. Rats consuming Ca-supplemented QPM products consumed more feed and gained more weight but had conversion ratios similar to those of their counterparts fed regular corn and sorghum. Among rats fed tortillas, QPM produced denser, stronger, longer, and thicker bones with more ash and Ca. Also, rats fed Ca-supplemented QPM products had the highest serum albumin levels, probably because of the improved dietary protein quality.

<u>Commercial production of QPM tortillas</u>. In 1986, a field study was conducted in a commercial plant in Mexico to compare the processing properties of a white QPM hybrid with a commercial Mexican corn [1]. White QPM and Mexican corn were cooked in 1% lime:water solution (based on corn weight) for 33 min and 40 min, respectively. Both cooked corns were steeped for 12 hours. Nixtamal was ground in a stone-grinder and then masa was extruded into thin discs in a "Celorio" sheeter. Masa discs were baked into tortillas in a triple pass gas-fired oven. The study concluded that QPM was easily cooked in alkali and processed into tortillas. The cooking characteristics were similar to available commercial Mexican corn

---

[1] Tellez-Giron, A. and Rooney, L.W. 1986. Unpublished data

varieties. QPM tortilla color, taste, flexibility, and other properties were similar or better than those found in Mexican corn tortillas.

Bressani et al.(1991) compared changes of major nutrients during village processing of QPM and regular corn cultivars. The changes in total protein, lipid, total dietary fiber, and ash contents from raw maize to tortilla were similar to those for regular maize previously reported by other workers. Rats fed diets made from raw grain, masa and tortilla made from QPM showed a weight gain, food intake and PER significantly greater than animals fed diets from regular corn and its processed products. Comparing different corn products, animals that consumed diets from masa and tortillas had a weight gain and PER significantly greater than animals fed raw corn diets. Concerning food intake, rats fed tortilla diets showed an intake significantly greater than those fed raw maize diets.

In 1990, the Cereal Quality Lab implemented a project to compare the processing properties of white and yellow QPM hybrids grown at Uvalde, Texas with a very good commercial yellow food-grade corn hybrid. These studies are reported below.

MATERIAL AND METHODS

Sample preparation. Three maize cultivars were used in this study: Asgrow 404Y, a commercial yellow food-grade corn with outstanding alkaline-cooking properties; a yellow quality protein maize (YQPM), which was a composite of grains from two experimental hybrids, F387 and 238 x 380; and white quality protein maize (WQPM), a composite of grains from three experimental white hybrids, 157 x 232, 176 x 230, and 173 x 232. The grain was grown at Uvalde, Texas in a replicated performance test under irrigation. The grain was air dried after harvesting, cleaned and stored in a freezer prior to processing.

Tortilla preparation and evaluation (pilot scale). Four kilograms of each corn variety were processed into tortillas (Fig.1). Grains were placed in perforated nylon bags and boiled with 50-L water and 166.6 g of lime in a steam kettle for the predetermined optimum cooking times. Optimum cooking time was determined in preliminary cooking trials described later. After boiling, steam was cut off, and the grain was steeped for 16 hr. The cooking liquor was discarded, and the grains were washed with running tap water to remove excess lime and pericarp tissue. Nixtamal was stone-ground (12in.-diameter lava stones) using a 20-hp commercial grinder (model CG, Casa Herrera Inc., Los Angeles, CA). Water was added during grinding at the rate of 0.6 L/min which increased masa moisture content to 55-56%. For all runs, the gap between the stone was adjusted to the same setting to produce a fine masa suitable for table tortillas. The masa was sheeted continuously and formed with a commercial sheeter/former (model CH4-STM, Superior Food Machinery, Inc., Pico Rivera, CA) into tortilla disks. Tortilla thickness was indirectly controlled by adjusting the sheeter to produce 35 ± 1 g masa disks, which were continuously baked into tortillas in a gas-fired oven with a three-tiers moving belt (model C0440), Superior Food Machinery). Tortillas were baked for 58 sec with an average temperature of 320, 290, and 370°C for the top, middle, and bottom tiers, respectively. Then tortillas were conveyed onto a three-stage cooling rack (model 3106-INF, Superior Food Machinery) with an average residence time of 60 sec.

Figure 1
Production of Table Tortilla

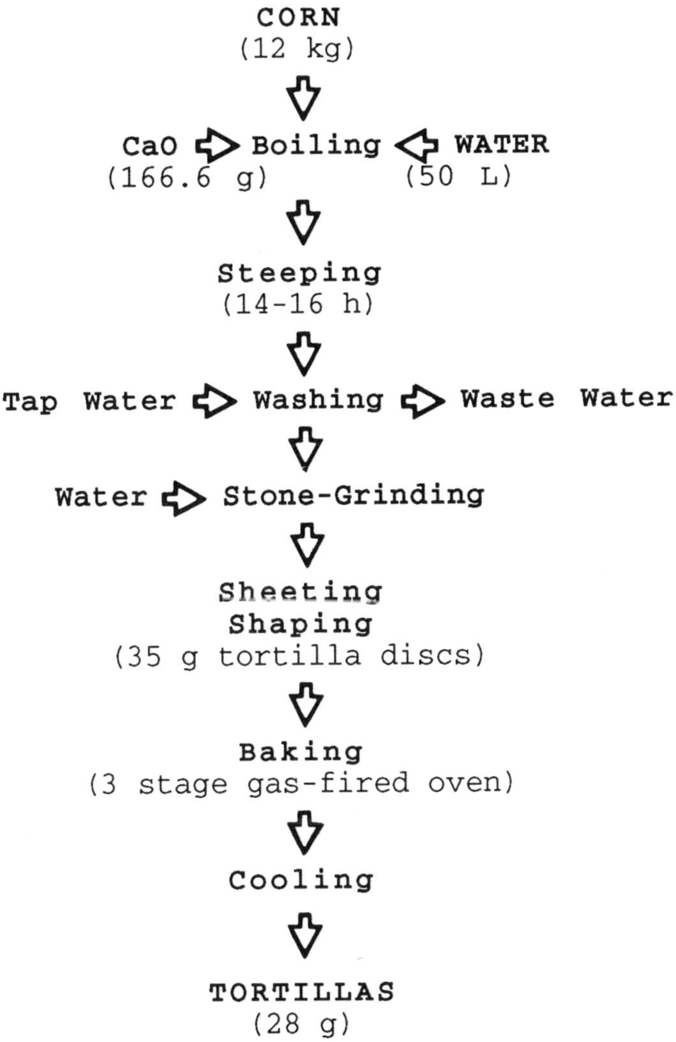

Tortillas were equilibrated at room temperature for about 10 min, weighed, and packed in low-density polyethylene bags.

Yields of nixtamal, masa, and tortillas for each hybrid were determined.

<u>Physical characteristics</u>. A 1000 kernel weight was counted with a semi-automatic batch type seed counter (Seedburo Equipment Co., Chicago, IL), and weighed.

Test weight or bulk density was determined with a Winchester bushel meter and expressed as pounds per bushel (or kg/hl).

Grain density was determined using a nitrogen comparison multipycnometer (model MUP-1S/N232, Quantachrome Corp., Syosset, NY) with a large cell.

Grain hardness (grams of matter removed per 100 g of whole grain) was measured by 10-min decortication of 40-g samples with a Tangential Abrasive Dehulling Device (TADD) equipped with an aluminum oxide abrasive disk and 8-hole base.

Objective color of masa and tortillas was determined using a HunterLab Tristimulus color meter (Model D25M-9; Hunter Ass.Lab. Inc., Reston, VA) (standard tiles: L= 76.21, a= -0.61, and b= 22.78). Color was expressed as a color solid according to the Hunter L a b system, where L is lightness (100) or darkness (0); +a is redness, -a is greenness; +b is yellowness, and -b is blueness.

Pericarp removal was evaluated subjectively after staining the lime-cooked kernels with eosine Y and methylene blue and washing with methanol as described by Serna-Saldivar et al.(1991).

Tortilla texture was determined subjectively by rolling a 1-in.wide tortilla strip around a 1 in-diameter dowel and grading

on a five-point scale (1= immediate complete
breakage and 5= no breakage during rolling).

Masa particle size was determined by
suspending 5 g of masa in 20-ml distilled
water and sieving through a set of 850 (US
standard sieve No 20), 250 (No 60), and 150
(No 100) μm-sieves (Gomez, 1988). Fractions
retained on each sieve were dried and weighed.

<u>Chemical composition</u>. Moisture content
was determined by drying to a constant weight
in a forced air oven at 105°C (AACC 1986).
Protein (%N x 6.25) was analyzed by Kjeldahl
digestion (AACC 1986) and an automated
colorimetric assay for ammonia (Technicon
1978). Crude fat in the samples was measured
by ether extraction using Goldfish apparatus
and evaporation of the solvent (AACC 1986).
Ash in the samples was determined after
incinerating at 500°C (AACC 1986). Total
dietary fiber was determined by a gravimetric,
enzymatic method which includes starch
gelatinization and digestion by alpha amylase,
pepsin and pancreatin (Asp et al., 1983)

<u>Determination of dry matter losses, nixtamal
water uptake and optimum-cooking time</u>. Corn
(100 g) placed in a perforated nylon bag was
boiled in water:lime solution for 0, 15, 30,
and 45 min and steeped for 16 hr. Water (50 L)
and lime (166.6 g) were brought to boil in a
120-L steam kettle (model TDC/2-20, Groen
Div., Dover Corp., Elk Grove Village, IL) and
bags of corn were added. The temperature of
cooking was kept at 97-98°C. During steeping
the temperature decreased at a rate of
0.15°C/min. Moisture content of nixtamal was
determined according to Serna-Saldivar
(1988a). The bags were opened, and nixtamal
was washed for 30 sec, drained, weighed after
10 min of air exposure, returned to the bags

for drying in a convection oven at 100°C for
48 hr., cooled in a dessicator, and reweighed.
Moisture content (MC), dry matter losses
(DML), and water uptake (WU) were calculated.
The time required to cause the nixtamal to
reach 50% moisture (cooking time) was
calculated using linear reqression equations
determined for each sample of corn. This
moisture level provided acceptable handling
properties during grinding of nixtamal,
sheeting, and baking of masa.

RESULTS AND DISCUSSION

<u>Properties of corn</u>. QPM samples had
lower test weight, reduced kernel weight and
were softer than the hard, yellow, commercial
food corn hybrid (Table 3).
The QPM kernels were significantly
smaller than the Asgrow 404Y grains. WQPM had
lower density than the other two cultivars. We
believe the optimum properties for alkaline
processing of Texas grown corn are test weight
of at least 60 lbs/bu, minimum density of 1.30
g/cc, 1,000 kernel weight of at least 300 g
and hardness of 40. Grains with these
properties have more tolerance to mishandling
and overcooking and tend to have reduced dry
matter losses during cooking. Alkaline-cooking
and steeping removed most of the pericarp of
Asgrow 404Y, 75% of YQPM and 50% of WQPM
(Table 3). Thus, the two QPM samples did not
meet all of the requirements for an ideal
Texas food corn; but with the exception of
kernel weight and pericarp removal they were
very close. Asgrow 404Y grain has outstanding
pericarp removal and other alkaline-cooking
properties. Other commercial food corns used
for alkaline processings do not meet all the
criteria either, i.e. Pioneer 3192 has very
poor pericarp removal. Thus, the QPM samples
can be processed into good table tortillas by

altering the cooking conditions since they are
significantly smaller than usual food corns.
Ultimately, types with more easily removed
pericarp can be obtained through breeding and
selection of QPM hybrids.

Table 3
Physical Properties of Grains

Property	Asgrow 404	YQPM	WQPM	LSD[a]
Test Weight lb/bu (kg/hl)	61.5 (79.2)	60.0 (77.3)	60.0 (77.3)	0.5 (0.6)
Density g/cc	1.33	1.33	1.31	0.001
1000 Kernel weight (g)	336	284	288	7.4
Hardness (% Removal)[b]	37.1	42.4	41.8	0.3
Pericarp Removal[c]	1.0	2.0	3.0	0.3

[a] Least significant difference (P<0.05)

[b] Amount of material mechanically removed in
a TADD mill after 10 min milling. Higher
values indicate softer kernels.

[c] Pericarp removal was subjectively rated as
1 = all pericarp removed, and 5 = all
pericarp present. This was using a
standard cooking time of 20 min for all
the samples.

The reduced kernel size of QPM shortens
its optimum cooking time, however this tends
to cause greater retention of pericarp in the
nixtamal. Pericarp remnants are not a real
problem for table tortilla production because

the retained pericarp is a source of gums
which probably prevent or retard staling of
the tortilla. For chip production, complete
removal of pericarp is required to avoid
problems in sheeting and forming of chips.
Thus, longer cooking times for QPM will be
required to remove most of the pericarp, which
will increase dry matter losses during
processing.

<u>Chemical composition</u>. The protein
content did not change during alkaline
processing into tortillas (Table 4).

Table 4
Chemical Composition of Grains and Tortillas[a]

Product	Prot. (Nx6.25)	Fat (%)	Ash (%)	TDF[b] (%)	COH[c] (%)
Asgrow 404					
Grain	9.8 a	4.9 a	1.2 b	12.1 a	72.0 b
Tortilla	9.9 a	4.2 b	1.4 a	10.5 b	74.0 a
YQPM					
Grain	9.5 a	4.8 a	1.5 b	13.5 a	70.7 b
Tortilla	9.8 a	3.4 b	1.6 a	12.4 b	72.8 a
WQPM					
Grain	10.4 a	4.3 a	1.6 b	13.8 a	69.5 b
Tortilla	10.3 a	3.2 b	1.7 a	13.0 b	71.5 a

[a] All values are expressed on oven dry matter basis. Means with different letter within column and grains are statistically different at P<0.05.

[b] Total dietary fiber (determined according to Asp et al., 1983).

[c] Total carbohydrate content determined by difference.

While, fat content decreased slightly,
suggesting that some fat-rich tissues (i.e.
germ) were removed during processing.

Tortillas had more ash or mineral
content due to the calcium uptake during
alkaline-cooking. Total dietary fiber
decreased from grain to tortilla in all
cultivars. Some of the pericarp was peeled
away by the action of lime during cooking and
steeping (Gomez et al., 1989). Most of the
undigestible carbohydrates, i.e., cellulose,
hemicellulose, and lignins are located in the
pericarp (Bressani 1990) so retention of
pericarp causes enhanced levels of total
dietary fiber. Thus, the fiber content of the
QPM tortillas was higher than that of the
Asgrow 404Y tortillas.

Lime-cooking trials. QPM grains absorbed
more water more rapidly than Asgrow 404Y
during lime-cooking and steeping (Table 5).
Dry matter losses were similar for the three
cultivars at equivalent cooking times.
However, Asgrow 404Y required a much longer
cooking time (26.3 min) to reach 50% nixtamal
moisture content. Serna-Saldivar et al.(1991)
reported that nixtamal with 50% moisture
content had adequate machinability and
produced good quality tortillas. The smaller,
slightly softer QPM kernels required less than
4 min cooking to reach the optimum nixtamal
moisture. The short cooking time and only 50
to 75% pericarp removal reduced the dry matter
losses of QPM grains compared to Asgrow 404Y.
The differences in cooking requirements
between QPM and Asgrow 404Y were mainly due to
reduced kernel size and to a lesser extent the
slightly softer endosperm texture of the QPM
grains. Grain, nixtamal, masa, and tortillas
contained similar moisture contents (Table 6).

Table 5

Effect of Lime-Cooking on Nixtamal Moisture and Dry Matter Losses of Asgrow 404Y, Yellow QPM and White QPM.

Cooking Time	Asgrow 404	YQPM	WQPM	LSD[a]
0 min				
Moisture (%)	45.2	49.9	49.4	1.4
DML (%)	8.3	8.7	8.9	o.8
15 min				
Moisture (%)	47.2	51.9	52.9	4.3
DML (%)	9.1	9.0	10.6	1.2
30 min				
Moisture (%)	49.9	54.5	55.8	2.8
DML (%)	9.2	9.6	11.3	0.9
45 min				
Moisture (%)	54.6	58.8	58.6	3.8
DML (%)	11.5	10.8	11.5	1.4
Optimum Cooking Time (min)	26.3	3.2	2.0	--
DML at Optimum Cook. Time (%)	9.4	8.6	8.3	--

[a] Least significant difference (P<0.05)

Thus, similar yields of nixtamal, masa, and tortillas were obtained (Table 7). Nixtamals were stone-ground into fine masa suitable for table tortillas. The three corns have similar masa fractionation values (Table 8) and excellent masa machinability. The most important fraction is the one that passes through the 150 μm screen because it is mainly composed of free native, annealed, and partially gelatinized starch granules

which are required to produce desirable
properties of masa.

Table 6
Moisture Contents of Asgrow 404Y and QPM
grains.

Product	Asgrow 404	YQPM	WQPM
Raw Grain	12.3±0.2	11.4±0.0	11.8±0.0
Nixtamal	51.4±0.4	53.2±0.2	52.7±0.4
Masa	57.2±0.3	57.2±0.5	58.7±0.4
Tortilla	46.2±0.5	46.0±0.3	46.8±0.6

Values are means ± standard deviation

Table 7
Yields of Nixtamal, Masa, and Tortilla from
Asgrow 404Y, and QPM grains.

Yield	Asgrow 404	YQPM	WQPM	LSD[a]
Nixtamal/Grain	1.86	1.95	2.02	NS
Masa/Nixtamal	2.12	2.18	2.29	NS
Tortilla/Grain	1.66	1.72	1.78	NS
Tortilla/Masa	0.78	0.78	0.78	NS
N° Tortilla / Kg Grain	63.0	64.0	65.0	NS

[a]NS: No significant difference (P<0.05).

As expected, WQPM masa and tortilla was
lighter in color than the products from

Asgrow 404Y and YQPM (Table 9a and 9b). The overall color index, E, is similar among masas and among tortillas.

All fresh tortillas had excellent rollability, taste and flavor. Tortillas still had acceptable rollability even after three days of storage at room temperature in sealed polyethylene bags.

Table 8
Masa Fractionation[a]

Fraction	Asgrow 404Y	YQPM	WQPM
Coarse (>250 μm)	22.2±1.0	19.9±1.2	18.0±1.4
Interm. (>150 μm)	3.8±0.2	2.7±0.3	3.3±0.3
Fine (<150 μm)	58.6±2.2	59.2±1.7	57.5±2.2
Soluble Solids	7.7±0.2	7.0±0.5	7.4±0.4

[a]Values are means ± standard deviation.

Table 9 a
Color (L, a, b, and E values) of Masa[a]

Color Value	Asgrow 404Y	YQPM	WQPM
L	70.3±0.2	72.5±0.3	74.9±0.1
a	2.34±0.02	0.36±0.02	-0.65±0.0
b	29.2±0.3	28.0±0.2	17.5±0.3
E	76.2±0.3	77.7±0.2	77.0±0.2

Values are means ± standard deviation.
[a] L= lightness (100= white, 0= black); a= red (+100), -a= green (-80); b= yellow (+70), -b= blue (-80); E= color index= $(L^2 +a^2 +b^2)^{1/2}$

Table 9 b
Color (L, a, b, and E values) of Tortilla[a]

CV[b]	Asgrow 404Y	YQPM	WQPM
L	65.6±0.3	63.6±0.2	67.9±0.2
a	1.63±0.01	0.41±0.01	-1.09±0.01
b	26.5±0.2	26.5±0.2	18.0±0.2
E	70.7±0.2	68.9±0.3	70.3±0.3

Values are means ± standard deviation.
[a] L= lightness; a= red (+100), -a= green (-80); b= yellow (+70), -b= blue (-80); E= color index= $(L^2 +a^2 +b^2)^{1/2}$
(b) Color Values

CONCLUSIONS

WQPM and YQPM required shorter cooking times than the food grade corn because of their smaller kernel size and slightly softer endosperm texture. The characteristics of products from QPM were similar to corresponding products from the food grade corn. QPM produced highly acceptable tortillas with excellent flavor, rollability and color. In addition, the QPM tortillas retained greater amounts of dietary fiber because the pericarp was not completely removed. For production of chips, a longer cooking time may be required to remove the pericarp completely and greater dry matter losses could be expected. However, tortilla chips fried from QPM had excellent color, flavor and aroma with a shelf life equivalent to regular corn tortilla chips. We have never observed any off odors or flavors with fresh QPM tortillas or

tortilla chips. QPM can be used to produce
good quality alkaline cooked products, but
processing conditions must be adjusted because
of its smaller kernel size.

ACKNOWLEDGEMENT

We acknowledge partial support for this
research from the Snack Foods Association,
Alexandria, VA.

Contribution TA 30298 from the Texas
Agricultural Experiment Station, Texas A&M
University, College Station, TX 77843-2474.

LITERATURE CITED

AACC. 1986. Approved Methods of the
American Association of Cereal Chemists.
St. Paul, MN.

Asp, N.G., Johansson, C.G. Hallmar, H.,
and Siljestrom, M.J. 1983. A rapid
enzymatic assay of insoluble and soluble
dietary fiber. J. Agric. Food Chem.
31:476-482.

Bockholt, A.J. Current status of QPM
hybrids for the U.S. Presented at the
Quality Protein Maize (QPM) Symposium.
American Association of Cereal Chemists.
1990 Annual Meeting. October 14-18.
Dallas, TX.

Bressani, R. 1990. Chemistry, technology
and nutritional value of maize
tortillas. Food Rev. Int. 6(2):225-264.

Bressani, R., Benavidez, V., Acevedo, E.,
and Ortiz, M.A. 1991. Changes in
selected nutrient content and in protein
quality of common and quality-protein
maize during rural tortilla preparation.
Cereal Chem. 67(6):515-518.

Gevers, O.H. Progress in developing
  modified *opaque-2* corn hybrids with hard
  endosperm, high nutritional value and
  good agronomics. Presented at the
  Quality Protein Maize (QPM) Symposium.
  American Association of Cereal Chemists.
  1990 Annual Meeting. October 14-18.
  Dallas, TX.
Gomez, M.H. 1988. Ph.D. Dissertation.
  "Physicochemical characteristics of
  fresh masa from alkaline processed corn
  and sorghum and of corn dry masa flour".
  Texas A&M University. College Station,
  Texas.
Gomez, M.H., Mc Donough, C.M., Rooney,
  L.W. and Waniska R.D. 1989. Changes in
  corn and sorghum during nixtamalization
  and tortilla baking. J. Fd. Sci. 53:330-
  336.
Mertz, E.T., Bates,. L.S., and Nelson, O.E.
  1964. Mutant gene that changes protein
  composition and increases lysine content of
  maize endosperm. Science 145:279.
Ortega, E.I., Villegas, E., and Vasal, S.K.
  1986. A comparative study of protein changes
  in normal and quality protein maize during
  tortilla making. Cereal Chem. 63(5):446-451.
Serna-Saldivar, S.O., Almeida-Dominguez, H.D.,
  Gomez, M.H., Bockholt, A.J., and Rooney,
  L.W. 1991. Method to evaluate ease of
  pericarp removal on lime cooked corn
  kernels. Crop Sci. 31:842-844.
Serna-Saldivar, S.O., Gomez, M.H., and Rooney,
  L.W. 1990. Technology, chemistry and
  nutritional value of alkaline-cooked corn
  products. In: Advances in Cereal Science and
  Technology. Vol. X. American Association of
  Cereal Chemists. St. Paul, MN.
Serna-Saldivar, S.O., Rooney, L.W., and
  Greene, L.W. 1991. Effect of lime treatment
  on the availability of calcium in diets of
  tortillas and beans: Rat growth and balance
  studies. Cereal Chem. 68(6):565-570.

Serna-Saldivar, S.O., Rooney, L.W., and
    Greene, L.W. 1992. Effect of lime treatment
    on the availability of calcium in diets of
    tortillas and beans: Bone and plasma
    composition in rats. Cereal Chem. 69(1):78-
    81.
Sproule, A.M., Serna-Saldivar, S.O., Bockholt,
    A.J., Rooney, L.W. and Knabe, D.A. 1988.
    Nutritional evaluation of tortilla and
    tortilla chips from quality protein maize.
    Cereal Foods World 33:233.
Technicon. 1976. Technicon Auto Analyzer
    II. Industrial Method No 334-74A/A.
    Technicon Industrial System, Tarrytown,
    NY.
Valverde, V., Delgado, H.L., Belizan, J.M.,
    Martorell, R., Mejía-Pivaral, V., Bressani,
    R., Elías, L.G., Molina, M., and Klein,,
    R.E. 1983. The Patulul Project: Production,
    storage, acceptance and nutritional impact
    of *opaque-2* corns in Guatemala. Institute of
    Nutrition of Central America and Pananma,
    Guatemala.
Vasal, S.K., Villegas, E., Bjarnason, M.,
    Gelaw, B., and Goerts, P. 1980. Genetic
    modifiers and breeding strategies in
    developing hard endosperm opaque-2
    materials. Pages 37-73 in: Improvement of
    Quality Trials of Maize for Grain and Silage
    Use, W.G. Pollmer and R.H. Phipps, eds.
    Martinus Nijhoff Publishers. Amsterdan.
    Holland.
Vivas-Rodriguez, N.E., Serna-Saldivar, S.O.,
    Waniska, R.D., and Rooney, L.W. 1990. Effect
    of tortilla chip preparation on the protein
    fractions of quality protein maize, regular
    maize and sorghum. J. of Cereal Science
    (12):289-296.